W9-CAP-638

After the World Trade Center

After the
World Trade Center

RETHINKING NEW YORK CITY

MICHAEL SORKIN and **SHARON ZUKIN** EDITORS

ROUTLEDGE
New York and London

Permission to reprint images on pages 92 & 133 is gratefully acknowledged.

> Berenice Abbott. Radio Row, Cortland Street between Washington and Greenwich Streets, April 18, 1936. Reprinted by Permission of the Museum of the City of New York, Federal Arts Projects

> Berenice Abbott. Lebanon Restaurant, 88 Washington Street between Rector and Morris Streets, August 12, 1936. Reprinted by permission of the Museum of the City of New York, Federal Arts Project

The following excerpt was previously published. Permission to reprint granted by Basic Books, a division of Perseus Books Group.

> Darton, Eric. Excerpt from *Divided We Stand: A Biography of New York's World Trade Center*. Copyright © Basic Books 1999, pp. 118–19.

Published in 2002 by
Routledge
29 West 35th Street
New York, NY 10001

Routledge is an imprint of the Taylor and Francis Group.

Copyright © 2002 Michael Sorkin and Sharon Zukin

All rights reserved. No part of this book may be printed or utilized in any form or by any electronic, mechanical, or other means, now known or hereafter invented, including photocopying and recording, or any other information storage or retrieval system, without permission in writing from the publisher.

Library of Congress Cataloging-In-Publication Data is available from the Library of Congress.

Sorkin, Michael and Sharon Zukin
After the world trade center: rethinking New York City
ISBN: 0-415-93479-6 (hardback)

CONTENTS

MICHAEL SORKIN and SHARON ZUKIN

Introduction

THEIR ABSENCE IS INDELIBLE: the Twin Towers were landmarks, buildings you could not lose sight of no matter where you were. They told you which way to face when you wanted to walk downtown, were your first view of the city when driving in from New Jersey, and anchored the skyline when you were flying out of JFK. As you rode the D train over the East River into Manhattan after dark, they were fluorescent chessboards against the black night sky. Sometimes sinister, sometimes beautiful, sometimes just banal, they were icons of New York City—the best-known buildings in the world, the Everest of our urban Himalayas.

When we saw the smoke and flames streaming from the towers in the hour before they fell, none of us imagined the collapse to come. And no matter how often we saw the media replays, it remained hard to believe these buildings were mortal, let alone the instrument of the death of thousands. However numbed we were by the compulsive repetition, we still couldn't get enough of it, couldn't stop staring at the plume of smoke that marked the void for weeks.

Suddenly, New York's gorgeous mosaic was tiled in tombstones by this equal-opportunity mass murder. The third-generation Irish, Italian, and Jewish stockbrokers, the women who had worked their way up to executive assistants and vice presidents, the Indian and Pakistani computer experts, the Caribbean security guards, and the Mexican cooks: of 50,000 who worked in the World Trade Center, nearly 3,000 died there. And they remain incredibly present—in moving daily

obituaries in the *New York Times*, in the missing posters that still cover so
many walls, in the lists of the fallen outside our firehouses.

Forced into a citizenship of common loss, the city banded together
soberly, spontaneously, generously, and with moving and unaccustomed
civility. We lit candles in our doorways at dusk. We gathered in Union
Square, turning it into a shrine and memorial, layered with photos, handwritten messages, schoolchildren's drawings, expressions of sympathy and
sorrow from flight attendants who had been spared by the luck of the
draw. New York was ready for its close-up in those early days, and the rest
of the country responded movingly. One night, a girl called one of us
from a small town in Oregon. "Just dialed your number because I wanted
to tell someone in New York how sorry we are."

But the mist of emotion also concealed both the dimensions of the
loss and the gradual worsening of old problems. It is estimated that
95,000 jobs were lost as the result of the tragedy, but we had also lost
75,000 jobs in the previous year. City finances are teetering at the brink
of fiscal meltdown, and the mayor has prepared us for a new round of
cutbacks that we can ill afford. Our picture of the city as a place of toler-
ance and freedom, where poor people can get an education and rise in the
world, is more and more at risk.

New York, the country, and the world are in the midst of an economic
recession. A Republican president who got little support from New York
City voters is fighting a new kind of war, and hands are daily wrung
about the difficulty of extracting promised aid from Washington. A Republican governor who is mounting a reelection campaign has set up his
own pipelines to Wall Street and is counting on the imminent adoption
of a rebuilding plan with his name on it. And a new Republican mayor,
whose career has been made exclusively in the financial community, must
cope with radically diminished resources and raised expectations. All this
occurs amid a heightened sense of global risk. We already have new
deputy police commissioners in charge of intelligence and counterterror-
ism, drafted from the CIA.

What, then, should we rebuild? Should the site be left as a memorial
as many—including numerous survivors of the victims—urge? Oklahoma City and Berlin have been wracked by prolonged conflicts over the

shape and meaning of such memorials. Do we want a garrison city, barri-caded against future attacks? Lower Manhattan hasn't been one since the threat of a British invasion subsided in the early nineteenth century. Do we want another downtown corporate financial center? Wall Street has been propped up by public subsidies and urban renewal plans for years, including, most recently, a huge—if suddenly shaky—tax giveaway to the New York Stock Exchange to build a new headquarters across the street from its old one. The office market itself has dramatically softened, sug-gesting that new space downtown is unlikely to be produced by simple supply and demand.

Since the fifties, companies in Lower Manhattan have been moving their headquarters to cheaper, greener pastures in the suburbs and setting up back offices far away, from Brooklyn to Bangalore. The computer rev-olution of the eighties, with its opportunities for radical decentralization, accelerated the trend. And with the advent of electronic trading in the nineties, all financial markets have had the potential to "dematerialize," leaving Lower Manhattan as the historic cradle of New York and a cul-tural center but eviscerating its old logics of concentration. Clearly, new styles of density must be introduced, new kinds of mix. Most important, new voices must be heard.

Unfortunately, the public agency that holds the power to make deci-sions about the site, the Lower Manhattan Development Corporation (LMDC), is cast in the old Port Authority mold. A committee of the powerful with only one representative from the local residential commu-nity, it's chaired by a Republican deal-maker and former director of Gold-man Sachs (itself about to move its entire equity trading department to a new billion-dollar complex in New Jersey). This Robert Moses–style au-thority has been given huge powers of legal circumvention and freedom from democratic oversight. On the bright side, this reversion to cronyism and the rule of money has been offset by the formation of numerous new civic associations bent on studying the rebuilding issue and producing schemes for renewal. None, however, has any real authority. Though the LMDC declared an initial "listening" period, it has held no public meet-ings. What it actually hears remains to be seen.

Anyone who is seriously listening, however, will hear a loud cacophony. Every issue past our initial grief has aroused conflict. Control over the World Trade Center site by firefighters or the police, the lack of insurance companies willing to pay for injuries sustained in clearing the site, and the inability or unwillingness to protect both workers and residents from environmental damage provoked the earliest expressions of conflict. Then there is the thorny issue of the memorial. How much of the WTC site will be dedicated to "unproductive" use? What will be the final design? Who are the "stakeholders," officeholders, and financial authorities with the power to decide?

Not unexpectedly, there has been a rapid and unseemly return to business as usual by many. Encouraged by the media, architects and planners trotted after the ambulance, ready to try to get the biggest job of their careers, joined by politicians and developers eager to thump their chests and proclaim the importance of rebuilding immediately. Everywhere the bromide is retailed that to rebuild something bigger, taller, and better than ever is the only way to respond to the terrorists. Few seem to suggest that our "victory" can lie only in a consequence that is positive for all of us, not in reflexive machismo.

More recently, we have seen ugly conflicts over political correctness and moral worth. A memorial statue of the three firefighters who, Iwo Jima–like, raised the American flag at the smoldering site, was redesigned to honor the city's ethnic diversity, even though these firefighters were all white, like most of the Fire Department. The quick unveiling of a federal government program for compensating victims' families with emergency funds also became grotesque. Though government officials tried to achieve a rough sense of equity by reducing the amount of compensation to reflect payments from other sources—a schema that reduced payments to more affluent families—those affected complained bitterly. We were then caught up in the spectacle of public calculations of actuarial worth, in which the potential earning power of stockbrokers from New Jersey was weighed against that of restaurant workers from Queens, reducing the victims to accounting abstractions.

We cannot reclaim the World Trade Center site without respectfully addressing its many ghosts. Although world trade in one form or another has always shaped this part of the city, the purveyors of its merchandise and the public spaces of its markets and entrepots have changed dramatically over the years. The earlier ghosts of this place are also victims of transformations that, if they have not always been cataclysmic, have often been violent. From the 1920 bombing of the Morgan Bank to the displacement of the largely Arab community that once thrived on the Lower West Side, to the destruction of an intimate architectural texture by megascale construction, this part of the city has been contested space. Though the destruction of the Twin Towers has reformulated the terms of conflict for the foreseeable present, it does not change this history.

The World Trade Center was the eye of a needle through which global capital flowed, the seat of an empire. However anonymous they appeared, the Twin Towers were never benign, never just architecture. Recovering this site for the living city is, therefore, inescapably political. Political, to be sure, because New York now takes a place in the long line of cities that have been damaged or nearly destroyed by terrorist and military attacks stretching, through the recent past, from Hiroshima and Berlin to Sarajevo and Kabul. Political because of Wall Street's role as an epicenter of world capitalism. Political because of Manhattan's site at the nexus of finance and real estate development—the city's most important industry. Political because of the power-brokering that will determine Downtown's future development. Political because of the growing clash between hallowed ground and buildable space.

Our intention, as a group of urbanists who live and work in New York City, is to use September 11, 2001, as an opportunity to speculate broadly about the future of the city we all love. The shadow of the towers demands that we both reconsider the past and think hard about the future. This book seeks to make a collective statement of purpose and of hope. We want to speak up for the task of history, the responsibility of architecture, and the needs of the living city, the whole city. We do not want our critical faculties to be subverted by our sorrow; we do not want the rebuilding of what was to take the place of building what should be.

Above all, we want to open a discussion wider than we have seen so far. We freely admit we do not know what lies ahead. By filling these pages with questions, however, we widen the door through which unexpected answers might come.

When Bad Buildings Happen to Good People[1]

The weight of this sad time we must obey;
Speak what we feel, not what we ought to say.
                    —Edgar, at very end of *King Lear*

I LIVE AND WORK at the other end of town. The radio told me to turn on TV fast; I was just in time to see the second plane crash, and then the implosions. My first thought was "Oh my God, it's like my book!" I meant *All That Is Solid Melts into Air*, a book I wrote in the eighties about what it means to be modern.[2] Now *All That Is Solid* is a good book, and it's full of wrecks and ruins and early deaths. A moment later, I thought: What's wrong with me? Parts of buildings and parts of bodies are flying through the air (if you had TV on, there was no way not to see), and I put *my* ideas and *me* in the foreground? But soon after that, on the screen and then in the street, I heard people talk, and I saw they were doing just what I'd done: making enormous mythical constructions that would make the whole horrific event revolve around *them*. We were like needy sculptors rushing to produce instant replacements for the giant stabiles that had stood on World Trade Plaza. We threw up anything we could hide behind, to hide our panic,

[1] The earliest version of this piece was written for *Lingua Franca*, wonderful magazine, late victim of the blast.
[2] *All That Is Solid Melts into Air: The Experience of Modernity* (Penguin, 1988).

our helplessness, and our instant, boundless sense of guilt. Shrinks call this "survivor guilt." Whatever else lies ahead for us, we can be sure there'll be plenty of this.

When we mourn a loved one or a leader, we often feel we are going back to something very, very old. September 11, 2001, was something else. Its instantaneousness and enormity made it new: this year's model in death. All of a sudden, "many thousands gone" was no cliché. We who didn't lose loved ones at the WTC had to ask: Where do we start to mourn? There was no shortage of places to start. There were all the devastated families and survivors; the thousands of people who lost their jobs in an instant on the 11th; the thousands more whose jobs kept vanishing all through the fall. Many of those who lost jobs were kitchen, delivery, and janitorial workers who served the WTC's thousands of offices. The media did well in getting us close to these people. The Family Assistance Center at Pier 94 seems to have done well in securing social services for them. But many were illegal immigrants, terrified to ask anyone for anything. All through the fall, when we saw more and more ragged, distraught-looking men in the subway and on the street, looking like they didn't know where they were going, we could be pretty sure it wasn't our imagination.

Firehouse, East 13th Street, September 2001. Photo by Richard Rosen.

Or we could start our mourning at the firehouse. All over town, it was the fire companies that took the worst hits. Live or on TV, we could see them rush into the bombed buildings to save people like us, and often not come out again. New York's firehouses are situated directly on crowded streets, legacies of our nineteenth-century "walking city." We pass them all the time and walk on by. Thanks to school field trips, our kids know the men and their gorgeous engines better than we do. But this fall we couldn't pass by the photos of the dead: the handsome young guys, the chiefs who looked like (and often were) their fathers and grandfathers. And we couldn't pass by the survivors, who looked just like the guys in the pictures, and who hung around outside, staring shakily into the middle distance, as if waiting for their friends to come back. One day I passed a company where no one was killed. I said I was so glad they had all survived, and a fireman with moist eyes said, "We shouldn't have." It seems that when Building No. 1 collapsed, they pulled everyone out of Building No. 2. But if their company had got downtown a minute earlier, they'd have been sent into No. 1 and died there with their friends. Is there balm in Gilead for their pain and guilt? They put their lives on the line, and their friends went over the line, in the name of public service. Many New Yorkers had to ask: Are we as a public worth this service? Can we even imagine being one? Sometimes I could, then the idea slipped away. I thought that if we could imagine it, maybe then we could touch their hands and look into their eyes, and help them heal their wounds.

There has probably been more talk about the Fire Department in the last six months than at any time since its formation 150 years ago. The media overflowed with stories of their courage and heartbreak. There was a widespread sense of civilian guilt—that we had taken them for granted. I certainly plead guilty. A fireman student gave me an FDNY T-shirt twenty years ago, but somehow I never wore it. What bottomless drawer is it at the bottom of today? If it turns up, will I dare to put it on?

A persistent question on many people's minds has been: *What can we do for them?* One recurrent answer has been direct, immediate offers of sexual healing—even from very respectable women who might have been firemen's sisters. I remember seeing on a local news broadcast a fireman who said he'd been hugged and kissed more times since September 11 than in his whole life. But the trouble has been that posttraumatic stress didn't appear to be any easier to heal at Ground Zero than it was thirty years ago at Vietnam's China Beach. The *New York Times* had a rueful piece in early December about Nino's restaurant and its throng of women volunteers: "To Serve and Flirt Near Ground Zero" by Victoria Balfour. So many of the encounters ended in mutual sadness. (But it's said that many women left their cards.)

There is another, collective form of healing that needs to be done at the FDNY. Through the years, the department's one big problem was that the deep bonds of family and networks of friendship that made it so humanly strong also made it impermeable, resistant to any initiative for civil rights. But so many were lost on the 11th, and the fire companies must be rebuilt fast. This will probably not be possible without including large numbers of blacks and women, who were excluded for so long. Their presence in the department can make it humanly strong in an entirely new way.

For weeks, and in some neighborhoods for months, the sense of loss has thickened the city air, merging, like the murky smells that drifted uptown, with incense from the candles on the streets, making it suddenly hard to breathe. The smell went away; then it came back. "Does it hurt?" "Only when I breathe."

For my wife and me, miles from the wreck, unable to get downtown till the first week's end, the most striking thing we saw all over was *the signs*. There was something special about these signs. At first they resembled the missing person signs you can see in every American city, with photo portraits and descriptive texts. We know that every year thousands of Americans walk out of their lives. But most do it on purpose—as suggested in the idiom "gone missing"—and most will eventually come back. If those signs scare us, our fear is that *this* man or woman or boy or girl has got so far out, all alone, that he or she won't be able to come back. But the new signs broke our hearts in a different way. For them, we only wished we could be afraid. We knew all too well these folks didn't "go" missing; we also knew they wouldn't be coming back. Working for the big companies in the WTC, or for the police or fire department, they were integrated into collective life. Mounted together on a hundred walls, these men and women had a stronger collective identity after death than they ever had in life. And they created a collective identity for us, one we didn't want but couldn't shed: *survivors,* the survivors of mainland America's first great air raid. But the signs on the streets also gave off a surprising individuality. As we walked around through the West Village, near the river, we noticed the names of several people repeated on signs that had been mounted by different signers, containing different portraits, different texts (sometimes endearments), and different numbers to call; it gave us an intimation of the complexities and ironies of their real lives.[3] The signs were windows; they helped us see some people just a little—sometimes more than the signers intended. We couldn't bring them back to life, but at least we could get to know them as more than numbers of the dead.

All through the fall, people kept coming back to these signs. What was so poignant about them? The signs dramatized one of the central themes of modern democratic culture: *life stories.* Life stories were a crucial force in the culture of the New Deal and the Popular Front, a culture that insisted that every ordinary person's life had meaning and power. Think of Joan Blondell singing "The Forgotten Man," or James Agee and Walker Evans's *Let Us Now Praise Famous Men,* or all the giant all-inclusive 1930s murals of ordinary people living their lives. But the ontology of the Popular Front was unfolded just after World War I, with the publication of James Joyce's novel *Ulysses.* Plenty of us cheered in 2000 when the New York Public Library judged *Ulysses* the greatest book of the twentieth century—so great, the library said, because it goes so deep into the life of a deeply ordinary man and his wife, and shows us how heroically extraordinary their ordinary life can be. Joyce's fascist ex-friend, the poet and painter Wyndham Lewis, condemned *Ulysses* as a sin against the spirit of the avant-garde; he mocked the book for its "Plainmanism."

The signs we saw downtown—with names like Ciccone, Lim, Murphy, Rasweiler, Singh, Morgan, Barbosa, Sofi, Vasquez, Pascual, Gambale, Draginsky, Bennett, Gjonbalaj, Vale, Alger, Holmes, and so many more—were triumphs of Plainmanism. The firemen rushing up the stairs were plain men very like the office workers rushing down. Together they

---

[3] On the inwardness and "narrative drive" of the creators of these signs, see Vivian Gornick's beautiful piece, "Why the Posters Haunt Us Still," *New York Times,* September 23, 2001.

give us some sense of the depth of being out there, in people we pass by on the street or rub up against in the subway every day. The diversity and contrasts of the names highlight America's most attractive quality, its "transnational," global inclusiveness, its openness to what Sly Stone called "everyday people" from everywhere. Some papers and television programs carried it on, broadcasting short lives.

The talk wasn't all Joycean in subtlety and depth. There was a tendency not to speak ill of the dead, and an inertia that bore us toward a "Lives of the Saints." I thought I would throw up if I had to hear about one more Little League coach. (Didn't they know how many of those parent coaches are pure poison?) Some informants offered details that made the dead real: this one hated his job, and had sued his company; that one was a lousy husband, and used malfunctions at the WTC as alibis—but still she'd take him back any day alive; another, a security guard, had carried on simultaneous affairs on floors 28 and 45—all dead; still another was a loan shark to five whole floors; so it went. Ironically, the more we heard embarrassing details, the more the victims came to life, and the more we missed them. In the dread light of those fires, all life seemed so sweet. I hope the great pain doesn't make us numb, and the empathy lasts; I hope it gets extended to those who have made the signs, the survivors, people who are here trying to live now.

It's a lot harder to feel empathy for those buildings. The earliest epitaphs for the towers were of the don't-speak-ill-of-the-dead variety. The Discovery Channel did a show on the buildings, hosted by John Hockenberry, an NPR commentator I used to admire. "Everything that is best in America," he said, "was embodied in these buildings." I felt America's enemies could say nothing more insulting about us than this compliment. By now, if we "speak what we feel, not what we ought to say," we should be able to face the fact that they were the most hated buildings in town. They were brutal and overbearing, designed on the scale of monuments to some modern Ozymandias. They were expressions of an urbanism that disdained the city and its people.[4] They loomed over Downtown and

---

[4] This story is told brilliantly in Eric Darton's critical history, *Divided We Stand: A Biography of New York's World Trade Center* (Basic Books, 1999).

blotted out the sky. (I admit, for the few with a view, they were glorious at sunrise and sunset, for about a minute apiece.)

The Twin Towers were purposely isolated from the downtown street system, and designed to fit Le Corbusier's dictum "*We must kill the street.*" Compare them to the Empire State and Chrysler buildings, located on main streets in the middle of city life. They evoked Corbusier's Platonic ideal of a modern building, "the Tower in the Park," only guess what? There was no park. The WTC was bedecked with noble language about global unity, but its real life, and the changes it went through, belonged to the swampy history of Manhattan real estate deals. It had plenty of "public spaces," but they were remarkably unfriendly to real people: too cramped and broken up, like the underground shopping arcade; too vast and void, like the windswept outdoor plaza, whose main achievement was to prove to the world that you can construct a desert in the midst of a metropolis.

I only wish I had a dollar for every New Yorker who has ever wished those buildings would disappear. And yet, and yet! "But Daddy," my seven-year-old son asked me late in September, "if you wanted those buildings to vanish, why did you get so sad when they did?" Because in the real world, when buildings disappear, they take innocent people with them. That's one of the ways we can recognize the real world.

Rereading this, I see how my petulant trash-talk dates me: it places me in a generation that can remember life BWTC. People who were too young to remember, who took their existence for granted, enjoyed them as landmarks and felt in awe of their size. And after the bombing of 1993, many people came to feel their vulnerability, and took pity. (On this evolution of feeling, see the moving 1998 poem by David Lehman, "The World Trade Center.") I shared this feeling, but I don't think it should be confused with love.

In its form, the WTC was brutal; in its functioning, it was a drag. As the eighties turned into the nineties, more and more people came to live downtown, and the neighborhood developed many interesting cultural scenes. But the Port Authority, the WTC's owners, seemed entirely uninterested. Remember where Dorothy in *The Wizard of Oz* says, "I have a feeling we're not in Kansas anymore"? As far as the WTC's directors were concerned, their buildings could have been in Kansas. Some people blamed their unresponsiveness on their leaden private/public institutional structure. But in fact, enormous "public authorities" can create environments far more humanly appealing than the WTC's. One of them is just next door: the Battery Park City (BPC) luxury housing complex and the office towers of the World Financial Center (WFC).

The BPC/WFC development arose about a decade after the WTC, and it is still rising. When Mario Cuomo, then governor of New York, appointed Meyer (Sandy) Fruchter as the BPC's first chairman, he instructed him to "give it soul, Sandy"; he was too polite to mention other megacorporate developments without soul. The BPC/WFC includes skyscrapers half the height of the WTC but far more graceful; they gave the space an organic form, like foothills leading up to a mountain range. There is a Winter Garden with giant palm trees, grassy knolls, a Jewish Museum, several vest-pocket parks, a bike path, a ferry slip, a yacht basin, a long and winding esplanade, a gorgeously landscaped cove, a platform where you can almost reach out and touch the Statue of Liberty. My wife and I had our first hot date there. WTC-type institutional structures can generate dramatic, romantic, and entertaining public spaces where huge city crowds can feel at home. But not at the WTC! It seemed so perverse. On any nice weekend over the last decade, the BPC/WFC was packed, the WTC virtually deserted—except for the great bank of escalators leading to the New Jersey PATH train. For years, the Port Authority kept the place as bleak as a bunker. Then, last July, it gave it up and leased it to developer Larry Silverstein for the next 99 years. Why did the PA hold on for so long, and why did it finally throw its hands up? The men and women who signed the lease went down on the 11th, so we'll never know.

For Jews who marked Yom Kippur, the Day of Atonement, on September 27, past and present seemed weirdly to collapse into each other. One of the day's central themes is trying to stay human in the aftermath of mass murder. We read, meditate on, and talk about the destruction of Jerusalem. This is an event that has really happened in history more than once, and could happen again, but it is also a symbol of the incommensurable horror that comes with the territory of being human. One prophetic text we read on this day is Isaiah 58. The author of this book, who is conventionally called "Second Isaiah" (some scholars say "Third

Isaiah," but I will settle for second), lived about two hundred years after the prophet Isaiah, during the Jews' Babylonian exile. He addressed their hope to return home:

> . . . your ancient ruins shall be rebuilt;
> You shall raise up the foundations of many generations;
> You shall be called the repairer of the breach,
> The restorer of streets to dwell in.

But God would support the project of rebuilding only if the Jewish elite fulfilled certain tasks. These tasks are not ritual, but ethical, directed not to God, but to other people:

> To loose the bonds of wickedness,
> To undo the heavy burdens,
> To let the oppressed go free,
> And to break every yoke. . . .
> . . . to share your bread with the hungry,
> And bring the homeless poor into your house;
> When you see the naked, to cover him,
> And not to hide yourself from your own flesh. . . .
> Then shall your light break forth like the dawn. . . .
> If you pour yourself out for the hungry,
> And satisfy the desire of the afflicted,
> Then shall your light rise in darkness. . . .
> And you shall be like a watered garden. . . .

If you want the power to rebuild, you need to share your wealth and your resources—your food, your clothes, even your homes—with those less fortunate than you. But beyond this, you have to learn new structures of feeling: recognize people less fortunate than you as "your own flesh"; don't "hide yourself" from people any more, but "pour yourself out." Only then can the city rise again. At least that was how we read it on this Yom Kippur.

Will Larry Silverstein "pour himself out" and share the big house that real estate economics has placed in his hands? With billions of dollars of disaster relief at stake, I expect he will for a while. Before he can exploit the property as his private cash machine, he will have to treat it as public property for a time. But what then? Ideas seem polarized between those who want to freeze the site in an agony of unending nothingness and those who just want to get business going as if nothing had happened. My feelings are mixed. I want life to go on. I know I'll feel better when Downtown fills up once more with all those limos honking at each other, and all those *Sex and the City* characters swaggering up and down the streets. But I don't want all those good people who were trapped in those bad buildings to be trapped again in our need to forget the worst. We need some way to keep those ruins alive, ruins that in some mysterious way were greater than their source. We need a memorial that can capture the vividness, spontaneity, and "narrative drive" of those signs. I want us to show New York's power not only to remember but to represent, in ways the world won't forget, the bond between the living and the dead.

But even as we remember, we need to resist the undertow, the after-shock, the time bomb of survivor guilt. The people who died came to downtown New York and put up with the dirt and noise and high prices and high rents and free-floating nastiness because they wanted to live and to make a mark in the world. If there is a bond between us and them, one thing we owe them is to be where we are, in the world, and stay more alive than ever.

New Yorkers need time to sort it all out. We should try to open fo-rums where people can say what they think and feel. We need a world-wide competition. Those first buildings were lousy, but there was something grand and inspiring in the global vision that underlaid them. That grandeur is what led many younger people to adopt them as sym-bols of the city. But maybe we can symbolize New York in ways that are more imaginative, playful, and humanly sensitive next time around.

One thing that may help us get it right is a chance to *participate*. This idea has many meanings. One is that we need to *talk*. New Yorkers are fa-mous for being big talkers, remember? "The weight of this sad time" has made many of us unusually sensitive to people very different from our-selves. America says it loves us now, but as the *Daily News* headline said it in the seventies, "Ford to City: Drop Dead," wasn't so long ago. We need to "speak what we feel, not what we ought to say." It's distressing, months after September 11, that people want so many different things—basically, as usually happens in New York, they all want exactly what they wanted before—and no overarching vision has emerged to bring them together. Maybe serious public dialogue will yield a shared vision of what we should build on the ruins? Maybe it won't; but we'll be happier if we know.

I even have a place for this rendezvous: Union Square. When Parks Commissioner Robert Moses closed the Speakers' Corner during World War II, on the grounds that somebody might say something damaging to national security, the suppression of talk started half a century of decline. The city rebuilt the square in the late 1990s. But it didn't really catch on as a public space—until September 11. Then, abruptly, it was flooded with candles, flowers, missing person signs, poems and drawings. Some art students unrolled a scroll of paper three feet wide and several hundred feet long. A great assembly of people gathered round the scroll, and wrote radically contradictory messages and meditations. Overnight, Union Square became the city's most exciting public space: a small-town Fourth of July party combined with a 1970s be-in. I knew it was too good to be true, and indeed, at the end of Week 2, the city shut it down. The mayor's office said that homeless people had begun to colonize the square (indeed they had), and that the city's campaign against the homeless trumped everything. Fences were put up to keep homeless people out, and of course they also kept out everyone else. I heard that some of the kids tried to hang on, on asserting the right to be a public there. Our new mayor may not share Rudolph Giuliani's hostility to live public space. In any case, it's nice to know that the space is there.

There was another sense of participation that would have made a lot of New Yorkers happy: to *schlep,* to get down and dirty, to help clear the endless rubble on the WTC site and open up the space. The task sounds like one of the mythical labors of some Hercules or Prometheus. But there was a time when a great many people would have been thrilled to offer their bodies, to exhaust themselves, to help the city come back to life. It could have brought people together as citizens, as a public. It would have been "urban renewal"—or maybe I should have said "community service"—for real. I can't imagine how it could have been worked out. But I know it's happened before at many points in history, after floods and fires and earthquakes and civil wars. I wonder if some of our business executives, whose brilliant creativity we are always hearing about, and who love to talk about New York's global grandeur and glory, couldn't have helped the city organize a job like this.

Though my suggestion for public participation in the largest public works project of our time—what they called the "cleanup job"—is no longer as relevant as it was when I first put these thoughts to paper in September, when the fires were still burning, it still raises an overwhelmingly important question for the future. How can New Yorkers participate in the changing life of the city? Talking in public space is good, but coming together for public work would be a better reason—the best reason I've heard in a while—to get up in the morning and go downtown.

Our World Trade Center

Now that the World Trade Center has been destroyed, and the sixteen-acre platform on which it stood has been forcibly flattened into the small-scale, almost medieval, street plan of Lower Manhattan, we can grasp more clearly what the place once meant to us and what it could mean to New Yorkers in the future. Ugly, awkward, functional—like the city itself—the Twin Towers made their great impression by sheer arrogance. They took over the skyline, staking their claim not only as an iconic image of New York but as *the* iconic image of what a modern city should aspire to be: the biggest, the mightiest, the imperial center. Once we gazed upon this site as a landscape of power, but since September 11, we have viewed it in sorrow—as if it holds both the dark side of grandeur and our unspoken fears of decline.

"This is Wall Street, still the financial capital of the world," the conductor on a Brooklyn-bound subway train announces after 9/11, as the cars shuffle through a nearly empty station. His optimism cheers us, for the nearby "Cortlandt Street–WTC" is now a moot destination. Closed forever, its ceiling temporarily propped up by wooden crossbeams, a large American flag suspended vertically at the midpoint, this station is where thousands of people jammed the platforms every morning on their way to work. Most of the office workers, stockbrokers, computer programmers, sales clerks, and messengers whom I used to see crowding through the turnstiles are gone, and we who remain on this line don't know whether they have just shifted to another route for their own convenience, moved to Midtown or Jersey City with their firm, or gone missing in the debris underground. The drivers slow the train when they pass through the empty station—in the first days after 9/11, the sadness in the car was palpable at these moments—and I can't help but think of the empty U-bahn stations in East Berlin that were boarded up for all those years after the Wall was built in 1961. Just as those ghost stations bore witness to Berlin's division between warring ideologies, so Lower Manhattan is now a site of conflict between two hostile regimes: the regimes of memory and of money.

A regime of memory must be organized around a central event or victim. But where do we draw history's fine line: whom and what do we remember? The families of the terrorists' victims and the survivors of the attack want to commemorate those who died on 9/11. In their desire for an instant memorial, they follow the example set by the nearest and dearest relatives of those who have perished in other recent disasters, especially in the bombing of the federal building in Oklahoma City. We feel the families' pain not only because their loss was so colossal, so sudden, and so senseless, but because we, too, might have died if our paths had taken us to Lower Manhattan that day.

But some of us also remember the World Trade Center's more distant past. We want to commemorate those people and stores, wharves and markets, that were fixtures on the Lower West Side before the World Trade Center pushed them out and away. We do not just mourn the victims of terror; we mourn an older city, a bustling and gritty urban center that didn't have chain stores or welfare reform or companies that do business just as easily from New Jersey, Trinidad, or Hyderabad as they do from Lower Manhattan. Our memory of this city is both evocative and selective. It resonates with the black-and-white photographs of street scenes taken by Paul Strand and Berenice Abbott. But it edits out Robert Moses and Austin Tobin, Tammany Hall, and David and Nelson Rockefeller, each of whom was memorialized in the giant construction scheme masquerading as an urban renewal project that the Twin Towers represented.

In the last quarter-century, we have surrounded our cities with memorials. A generation of Holocaust memorials, not only in Berlin and

Auschwitz but also in Washington, D.C., and Lower Manhattan, are hallowed sites in a tourist's itinerary of collective pain. The Vietnam Memorial on the Mall, the Civil War battlefields and battle reenactments: we have plenty of inspiration for revisiting grief. Closer to home, we work through our grief by recalling the hundreds of spontaneous memorials that we ourselves erected after 9/11, when we transformed the entire city into a regime of memory. Shrines made of flaming candles dripping wax, fading floral bouquets, hand-lettered signs, and crayoned drawings filled Union Square Park for weeks. At the park's entrance, the statue of George Washington mounted on a horse was covered with wreaths of flowers, and love and peace signs were painted as white graffiti on its base. Following George Washington's pointing finger, we peer into the sky downtown, toward where we used to be able to see the tops of the towers. Our memory completes their outline in our minds. We find comfort in the old-fashioned buildings that remain: the Empire State Building on 34th Street, the loft buildings of Tribeca and SoHo, and the ornate, early twentieth-century skyscrapers of Wall Street, whose arrogance so perturbed Henry James.

A regime of memory seeks permanence—yet this is the very quality that James complained was absent in New York City. Skyscrapers have always been built for love of money, not for permanence, or public purpose, or art. No one could confuse the World Trade Center with the Palazzo Pubblico or New York with Siena, Athens, or the Valley of the Kings near the Nile. Even if remembrance is the sacred religion of our time, our profane, daily culture—as James well knew—is based on money.

Though we can summon memory by building monuments, who controls the even more elusive regime of money? The Twin Towers didn't belong to the former regime—they belonged to three concentric circuits of money and power, which tightened in a narrow noose around the sixteen-acre site on the Lower West Side.

The towers belonged, first, to a global circuit of capital flows, where money—or its abstract symbols—passed through national stock exchanges, multinational banks, and global trading firms just as their local employees passed through the turnstiles at Cortlandt Street. Some of these firms were located in the World Trade Center, more were clustered in the financial district around the towers, and many more were a phone call away in Midtown, Brooklyn, or New Jersey. Their business was the world's business—raising funds, guaranteeing credit, moving money from one portfolio to another. It is not surprising that several of the employees who died on 9/11 were corporate travel planners, for the emblem of this circuit is travel.

But the World Trade Center also took part in a circuit of money and power that was centered in Lower Manhattan, a circuit that has deeply changed in recent times. Wall Street—the historic financial district—claimed to be the geographical capital of capital, yet how metaphorical "the street" had become during the past few years, when Merrill Lynch and Dreyfus opened branch offices in local banks and shopping malls all over the country, and mutual funds and electronic trading systems began to control a larger share of small investors' savings. In fact, Wall Street as a real geographical space, a functioning center of capital, has existed on artificial life-support systems since the fifties. The Chase Bank's plans for urban redevelopment, the Port Authority's giant construction project at the WTC, the outlying apartment houses and public schools of Battery Park City: all would have collapsed after the stock market crash of 1987, had it not been for the unexpected diversity brought by small-scale cultural industries such as TV and film production, graphic arts studios, and restaurants, and by the aesthetic draw of Hudson River Park. If not for this infusion of new people and activities, the whole district would likely look, at best, like Chambers Street, with its discount stores, XXX-rated video shops, and fast-food restaurants. I could see the emblem of this circuit changing from finance to culture in the nineties, when I overheard a man dressed in a business suit say to his friend in the lobby of an office building on 43rd Street, "I went down to Wall Street yesterday—I had a great dinner in Tribeca."

And yet the Twin Towers owed their brief life directly to the third circuit of money and power, the one based on the marriage of global finance and local real estate interests. In contrast to other cities, New York's main business is and always has been real estate development. Developers are the engines of the city's endless cycles of boom and bust; they abandon the old and make a fetish of the new, and there are always fresh ones who can be persuaded to try to make their fortune by promising to rebuild Manhattan. Unfortunately, politicians believe their promises. And the World Trade Center was a prime example of how deeply indebted the public sector becomes to private-sector development. The World Trade Center merged the roles of public and private developer: the Port Authority—a public authority of New York and New Jersey—took the financial risks that private developers are supposed to take, while reaping few of the profits.

The shift from the Port Authority's ownership to Larry Silverstein's leasehold just before 9/11 signaled that a new boom for office building owners was under way. A small boom had been nurtured during the past few years by the Alliance for Downtown New York, Wall Street's own Business Improvement District. This boomlet promised, however, to steer development onto a somewhat different path, one that would capitalize on museums, walking tours, and residents who also work in the creative arts. It's hard to know whether this represented a newfound realism or was just a straw in the wind. Would any business group really give up control so Wall Street could return to being a real space instead of being the metaphorical capital of financial capital?

It's not easy to force the regime of money to conform to other priorities. Certainly now, with both the state and the city on the cusp of a new fiscal crisis, the regime of money is playing the card of realpolitik. I don't think they're bluffing. Republicans hold the White House, the State Capitol, and City Hall. Though 25 million square feet of office space were removed at one blow from Lower Manhattan, vacancy rates are alarmingly high. And corporations managed by a footloose elite won't prop up Wall Street forever. These institutions will not invest in Lower Manhattan without clamoring for big public subsidies, as the commodities and stock exchanges have already shown us, with their threats to move from the capital of hubris into the eager arms of humble New Jersey across the river. Even the saddest firm that was left bereft by the World Trade Center's demise—the stock and bond traders at Cantor Fitzgerald—has shown us how companies quietly shift their business strategies without thinking about the social costs. Improbably, in light of the great loss of lives the firm suffered on 9/11, Cantor Fitzgerald

Overlapping the regimes of memory and money: street banner, Broadway, February 2002.
Photo by Richard Rosen.

declared a fourth-quarter profit at the end of 2001, thanks to a strategy already in use of replacing human traders with an electronic system, and thanks to their ownership of eSpeed, a separate, electronic trading firm.

This is why I feel ambivalent about the juggernaut of public and private interests that is catapulting Lower Manhattan toward rebuilding offices for financial firms. Can we alter cycles of growth—or destruction—for the common good in a way that does not end in greed and selfishness? We have already heard too much ominous rumbling over money, from the reluctance of Swiss Re, the insurance company that holds the policy on the WTC, to pay rebuilding costs to Larry Silverstein, to the sudden withdrawal of policies against terrorist attacks by other insurance companies. And we have also seen too many turf wars over the available resources, from the victims' families' claims for unlimited reparations, despite a plan to share the benefits developed by federal emergency officials, to the scandal that enveloped the American Red Cross because of the way the organization allocated the money and blood it collected after 9/11.

Yet money is the city's lifeblood. While the Lower Manhattan Development Corporation promises to promote economic development by rebuilding office space downtown, the city urgently needs to rebuild jobs. But in which economic activities, and at what levels, would these jobs be? Who will get the money? The $250,000-a-year stockbrokers and bond traders who threaten to move to New Jersey, or the welfare mothers who take care of the city's parks and have already been demoted to temporary workers at $7.83 an hour?

To be sure, the World Trade Center supported tens of thousands of workers at many different levels: bond traders who commuted from the suburbs, office managers and clerical assistants from the Bronx, Polish janitors and young data analysts from Brooklyn, Mexican cooks and Salvadoran delivery men from Queens. And then, in the nearby neighborhoods of Lower Manhattan, there were waiters who moved here from Fukien, garment workers from Hong Kong, and taxi drivers from India and even Afghanistan. But the prosperity of some never ensured the livelihood of all.

The regimes of memory and money overlap to some degree in the tourist economy—in the city's museums, theaters and cafes, and ethnic neighborhoods—where the performance of life is a salable commodity. But the many tourists who have come to New York in recent months come to witness the performance of death. An impromptu infrastructure—small but growing—has sprung up to cater to their needs. Street vendors near the viewing platforms of the WTC site at Broadway and Fulton streets sell American flag pins and picture postcards with views of the Twin Towers at twilight. You can pick up your free viewing ticket at South Street Seaport, file past the site, then buy an NYPD cap and an FDNY sweatshirt, without forgetting to take a snapshot to show the folks at home. But many New Yorkers feel uneasy about this sort of sightseeing. It doesn't memorialize the dead so much as it tries to make a connection between the dead and the living, with the "dead" being the media images of destruction we all saw on TV and the "living" taking the form of branded logos we can wear on our backs. Between the individual memories of horror and the trivial act of buying a souvenir, the city disappears.

And that, of course, is our worst fear.

To disappear by the wrecker's ball or the terrorist's bomb, in the blitzkrieg of war or with suddenly downgraded credit ratings on municipal bonds: we don't want to contemplate these terrible alternatives. Yet suddenly, at the World Trade Center site, we look them all in the face, and we slip back and forth between memory and money—between remembering the grandeur of being "the capital of the world" and detesting the arrogance of power.

No, we who live here don't want New York to disappear. Many of us don't really care about the regime of money, as long as we don't get the 2012 Olympics or two new baseball stadiums the city doesn't need, and the poorest New Yorkers don't wind up paying for a $4 billion municipal budget deficit while the wealthiest 1 percent get federal tax cuts. We do care about the regime of memory. And the finest memorial to 9/11 would be to use the destruction of the World Trade Center to understand people's ambitions in other parts of the world—and to understand our own ambitions, also.

What are our ambitions? We can't get back the willful innocence we shared with the rest of America; we never had real innocence—even in colonial New Amsterdam, they kicked out the Munsee Lenape, imported

Africans to do much of the work, and drank water from polluted wells. But we can capitalize on our history of self-invention. Let's not build offices in Lower Manhattan and pay companies to move there; and let's not downsize, but *reshape* New York.

Let's encourage creativity in both the arts and business by nurturing a supply of low-rent space. Let's prevent crime by creating jobs for ordinary people—light manufacturing jobs to make the things dreamed up by creative designers; communications jobs; building and supplying jobs. Let's enhance the comfort factor of the old—by making our old subway trains run on time, by renovating our old buildings, by putting humans back in customer-service jobs now badly performed by automation. Let's make the wealthy less visible: eliminate the celebrity photos and gossip columns of Page Six, limit the number of high-price designer boutiques. Vow to maintain a walkable, affordable shopping center like the Bronx's Arthur Avenue in every neighborhood. Let's get the City Council to work together with the Community Boards—not to scare away businesses, but to treat them as real anchors of the community.

We have seen how a city looks when large parts of it are destroyed. From the devastation of Kabul in our time to the recent history of the South Bronx, Bushwick, and Harlem—not omitting Newark, New Jersey, and Bridgeport, Connecticut—cities cry out to be rebuilt. But let's not rebuild in arrogance. We don't need more superblocks and mammoth centers, we need many, smaller centers. We need to rebuild a lower-scale Downtown where life hums and throbs on every block. This is what the World Trade Center has taught us.

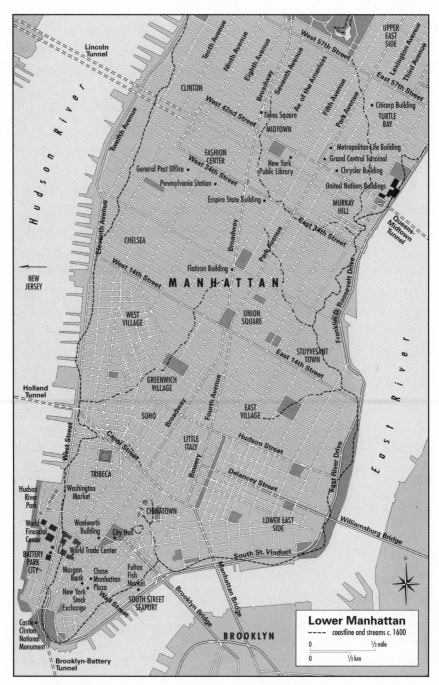

Manhattan's coastline, 1600 and the present.

Manhattan at War

THE TERRORIST OUTRAGES of September 11 serve as an unwelcome reminder that Lower Manhattan has abundant experience with the violence spawned by international trade and war. In 1625 the Dutch West India Company chose that site for a fur-trading post in New Netherland, "purchased" the entire island from the Lenape Indians, and began work on a massive fortress at the foot of Broadway (now the site of the Museum of the American Indian). The fort's location made military as well as commercial sense: it was the prime site from which to control traffic along both the East and Hudson rivers, and it would also guard New Amsterdam, the burgeoning settlement outside its walls, from attack by the Spanish (with whom the Netherlands had been fighting for decades). In the years that followed, the ebb and flow of world events would reveal new enemies and compel the residents of Manhattan to improve and extend their defenses. I want to suggest, indeed, that that sense of exposure, of precariousness, of vulnerability was central to their historical experience.

Because the fur trade failed to generate the profits it had hoped for, the West India Company devoted less attention to the security and development of New Amsterdam than to its extensive operations in Africa, Latin America, and the Caribbean. It never entirely gave up on the town, however, and the population climbed, by fits and starts, from perhaps a few hundred in the early 1630s to some 1,500 in the early 1660s. During those same years, two public works—the erection of the first municipal pier near the foot of Whitehall Street and the transformation of a sluggish creek into a shipping canal that sliced into town from the East River—promoted the gradual expansion of overseas trade, as local merchants sought markets in Europe and the West Indies for locally produced tobacco, grain, timber, and potash. Both the pier and the canal also helped fix New Amsterdam's orientation toward the East River waterfront and foreshadowed the increasingly ambitious construction projects that would transform the geography of Lower Manhattan over the next several hundred years.

A third project—the building of a wooden stockade across the island along the route of present-day Wall Street—bore witness to the shifting fortunes of the Netherlands in both America and Europe. In the 1630s and 1640s, English colonists began to move across the ill-defined borders dividing New England and New Netherland, occupying land claimed by the West India Company and defying efforts by the Dutch to drive them off. Peter Stuyvesant, director-general of the colony since 1647, tried to stem the tide through a combination of diplomacy and the organization of new Dutch towns on western Long Island. His efforts were complicated, however, by the looming crisis in Anglo-Dutch relations that followed Parliament's adoption of legislation that excluded the Dutch from trade with English overseas possessions. In 1653, with the two nations at war, word reached New Amsterdam that English forces were massing in Boston for an attack on Manhattan. Fearing the enemy would strike by land as well as by sea, Stuyvesant and the magistrates hastily stockaded the northern perimeter of the town and ringed it with a number of small breastworks. The danger passed when England and the Netherlands agreed to begin peace talks, but it was more evident than ever that the fate of Manhattan was inextricably tied to the outcome of conflicts orchestrated thousands of miles away by politicians and generals who would never set foot on it and might have had no clear idea of where it was.

That lesson was underscored a decade later, in 1664, when the English raided Dutch outposts in Africa and King Charles II authorized his brother, the Duke of York, to seize New Netherland. In August of that year, the duke's small fleet anchored in Gravesend Bay and dispatched an advance party of 450 soldiers and sailors to seize the ferry at Breuckelen, directly across the East River from New Amsterdam. Stuyvesant prepared to make a fight of it, but when the English promised generous terms for a peaceful capitulation, the townsfolk refused to back him up. On September 8, 1664, the West India Company's colors were struck from the fort, and the soldiers of its garrison marched down to the pier to board a ship for the long trip home. New Amsterdam and New Netherland were promptly named New York. Six months later, England and the Netherlands came to blows again.

When the Second Anglo-Dutch War finally ground to a halt three years later, the Dutch government agreed to let the English keep New York in exchange for Suriname (Dutch Guiana), whose slaves and sugar plantations were more highly valued by the West India Company. Dutch residents of the town continued to hope for a restoration of Dutch rule, however, and when the Third Anglo-Dutch war broke out, their hopes were rewarded. A huge Dutch fleet—twenty-odd warships and 1,600 fighting men—crossed the Atlantic in the summer of 1673 to attack English possessions in the Caribbean and along the coast of North America. The fleet reached New York at the end of July and ordered the surprised commander of the English garrison to surrender. When he stalled, the Dutch bombarded the fort and landed 600 marines on the Hudson River shore near the present site of Trinity Church. Cheered on by the Dutch populace, the marines advanced down Broadway and took the fort without firing a shot. New York now became New Orange in honor of the young Prince William of Orange, whose heroics had helped the Dutch win battle after battle in Europe. But New Orange was a chimera created by the vagaries of war. When the conflict ended less than a year later, the Dutch gave it back.

New York remained securely in English hands for another century. Its population climbed doggedly from an estimated 3,000 in 1680 to 7,000 in 1720, and to 18,000 in 1760—bigger than Boston (15,800) and second only to Philadelphia (23,800), the largest city in British North America. New residential and commercial construction meanwhile pushed the built-up area of town further and further up the island, past Wall Street (the crumbling old stockade came down in 1699) toward the Common (today's City Hall Park). By 1775, New York was home to some 25,000 people and could boast of a college, numerous churches, a synagogue, a new city hall, a theater, and busy public markets. Lower Broadway, well removed from the tumultuous East River waterfront, was now the town's most fashionable neighborhood, thanks in no small part to the allure of Bowling Green, a leafy little retreat laid out in 1732.

The very topography of the city was changing as well. To meet the incessant demand for building lots, municipal authorities ran new streets through orchards and pastures, drained swamps, buried streams, and leveled hills. Back in the 1670s, the old Dutch canal had been filled in to create an usually wide thoroughfare—dubbed, appropriately, Broad Street—which soon became one of the town's principal commercial arteries. At its foot, the city built a new stone pier, much larger than its predecessor and protected by two great moles, or breakwaters, that arced out into the East River. But even this impressive structure could not accommodate the swelling traffic in and out of the port, so the city encouraged merchants and shipowners to construct their own wharves by filling in and building on tide-lots (a solution facilitated by a new municipal charter, which gave the municipal corporation title to all land lying under water between low tide and high tide). During the eighteenth century, as a result, the East River waterfront began to creep, block by block, beyond the island's original shoreline—a process that continued at an even faster pace during the nineteenth century. The same thing occurred on the West Side. Portions of Greenwich and Washington streets were laid out and filled as early as the 1720s, and by mid-century new docks and warehouses had appeared along the Hudson River between Trinity Church and St. Paul's. (The expansion of Lower Manhattan into the Hudson slowed after West Street opened in 1830. It resumed in the late 1960s, when one million cubic yards of earth removed from the World Trade Center site were used as fill for the construction of Battery Park City.)

What did not change in the eighteenth century was New York's intimate and potentially dangerous links to the rest of the world; indeed, as the city became an integral component of the emerging British Empire, its residents were arguably more conscious than ever that their prosperity and peace of mind depended on events far removed from Manhattan. Everyone understood, for example, how the ballooning slave populations

of the West Indian plantations—a consequence, in turn, of the rising demand for sugar in Britain—had become a cornerstone of the local economy by the end of the seventeenth century. The need to feed and clothe
hundreds of thousands of enslaved Africans created lucrative markets for
the surplus foodstuffs and raw materials produced by area farmers and
put money in the pockets of the merchants who transported those foodstuffs and raw materials down to the islands. Money in the pockets of
merchants meant employment for the tradesmen who manufactured
rope, sails, and barrels for the growing number of ships working out of
the city as well as wages for vast numbers of sailors, cartmen, tavern keep-
ers, and casual laborers. The extent of New York's dependence on its connection to the West Indies was emphasized by the seasonal rhythms of
municipal business—feverish between November and January, as merchants scrambled to get down to the islands in time to take delivery of the
new sugar crop, then again between April and June, as they raced back to
port to escape hot-weather diseases and hurricanes.[1]

Perhaps the most visible consequence of the West Indian connection,
however, was the extraordinary concentration of slaves within and around
the town. Although the West India Company had relied upon slave labor
for construction projects in New Amsterdam, it would not become indis-
pensable until after the beginning of the eighteenth century, when city
merchants took an increased interest in the slave trade and began import-
ing multitudes of slaves for sale to area residents. Between 1700 and
1775, some 7,400 slaves would arrive in the city—more, in other words,
than its entire population at the turn of the century—and by 1750
African-Americans made up roughly 20 percent of the population, the
highest concentration of slaves north of the Mason-Dixon Line. Merchants used slaves to fill out crews and toil on their docks, artisans used
slaves in their shops, and Long Island farmers used slaves to till their

---

[1] And frequently brought the diseases along with them, with devastating results. In
the summer of 1702, an epidemic of what was probably yellow fever claimed 580 lives,
roughly 10 percent of the population. Milder outbreaks occurred so often over the next
century that well-to-do residents made a point of leaving town during July and August. In
1798, however, the fever killed over 2,000 New Yorkers, close to 5 percent of the population.

fields and tend their herds. So insistent was this demand for slaves that as early as 1711 the city established a market for their purchase and lease at the foot of Wall Street.

For white New Yorkers, the size and broad distribution of their servile population—roughly half the households in the city held one or more slaves—proved a source of constant anxiety. Disobedient and defiant slaves were a fact of life that required constant tinkering with local ordinances and spawned frequent rumors of revolt, typically linked to lurid accounts of slave uprisings in the Caribbean. The city's first taste of open insurrection came in 1712, when a group of slaves set fire to buildings and ambushed residents who rushed to put out the flames; two dozen slaves were eventually hanged, burned at the stake, or broken on the wheel for their part in the business. In 1741, fears of another uprising swept the city and led to the executions of thirty blacks and four whites. Although the extent of the 1741 conspiracy remains a matter of dispute, the panic it caused was fueled by the outbreak of war in Europe and the departure of many soldiers from the garrison for an expedition to the Caribbean. That New York now lay exposed to its enemies deeply concerned white residents and did not escape the attention of blacks, who recognized that they had been handed a perfect opportunity to rise up against their masters.

Like the fear of slave revolts, war was also a fact of life in British New York, touching residents in myriad ways. Already twice conquered and keenly aware of the Anglo-French rivalry that now dominated European affairs, they spent their days in an almost constant state of alert and hemmed in by defensive works—the hulking fort that still occupied the southern tip of the island, gun emplacements scattered along both shores (one of which, built directly below the fort in 1693, would inspire locals to speak of the area as "the Battery"), and a new palisade with gates and blockhouses thrown across the width of Manhattan in 1745 when officials got word that the French were planning to descend on the city from Canada (it zigzagged along a route sandwiched between present-day Chambers and Canal streets). Then, too, because its harbor and location made the city an especially convenient station for both the army and the navy, residents must often have thought themselves under more or less continued military occupation—great warships riding at anchor in the harbor, throngs of sailors on shore leave prowling the streets in search of a good time, regiments of redcoats drilling on the Common. In 1755, to no one's surprise, New York became the headquarters of all British forces in North America, and the government initiated monthly packet service between Falmouth and Manhattan, confirmation of the city's unique importance in the empire.

War and preparations for war contributed significantly to the city's economy as well. Provisioning His Majesty's forces required gargantuan quantities of food, clothing, shoes, rum, horses, wagons, and other materiel that meant brisk business for local merchants and artisans, who typically greeted the outbreak of a new conflict or a new campaign as a boon. For those inclined to high-risk adventure, it afforded an opportunity to obtain letters of marque and go aprivateering in the West Indies. In fact, more privateers would operate out of New York during the eighteenth century than out of any other Atlantic port, and they returned home with prizes worth something like two million pounds sterling—an immense accession of wealth and the basis of more than one family fortune.

During the 1760s and 1770s, as American resistance to parliamentary taxation spiraled into revolution, New York's strategic value to the empire would make it a crucial military objective for both sides. Led by George Washington, the Continental Army, ten thousand strong, occupied the city in the spring of 1776, throwing barricades across streets, building forts, and erecting new shore batteries, certain in the conviction that the British would sooner or later attempt to recapture it so as to drive a wedge between the northern and southern wings of the rebellion. They did not have to wait long. At the end of June, lookouts on the Battery spotted the initial contingents of what proved to be the largest British expeditionary force before the twentieth century. All told, stationed on or near Staten Island were two great men-of-war and two dozen frigates mounting a combined 1,200 cannon, plus 400 transport ships, 32,000 soldiers, and 13,000 seamen. The British attack began at the end of August, when 15,000 redcoats crossed the Narrows to Long Island and occupied the flat southern portions of Kings County (equivalent to present-day Brooklyn). Their goal was Brooklyn Heights, directly across the East River from New York. On the morning of August 27, fortified by the addition of another 10,000 redcoats and 5,000 Hessians, the British assaulted American positions blocking the approaches to the Heights. By early afternoon they had driven the Americans from the field, killing perhaps 1,200 as well as capturing three generals and some ninety junior officers. Although Washington managed to get the demoralized remnants of his army back to Manhattan, New York's fate was sealed. In mid-October, the city fell for the third time in little more than a century.

New York lay under enemy occupation until the end of the war in 1783, enduring two catastrophic fires, one of which raged up the West Side of town and consumed Trinity Church, as well as the widespread destruction of both public and private property at the hands of His Majesty's forces. But once they left, a combination of factors enabled the city to rebound quickly: the aggressive exploitation of new markets by local merchants, an influx of newcomers equipped with both capital and connections, the organization of the first bank and stock exchange, and even the arrival of the Continental Congress, which demonstrated its confidence by moving into quarters in City Hall in 1785. The pace of New York's recovery can be read in the growth of its population. In 1790 the first federal census revealed that New York's population stood at a record 33,000—still smaller than that of Philadelphia, but rising more quickly. Only twenty years later, in 1810, there were more than 96,000 people living in the city, making it the largest in the United States. By 1830, its population exceeded 200,000.

Liberation, national independence, and spectacular demographic expansion did not, however, allow New Yorkers to feel significantly more secure than their colonial predecessors. During the Napoleonic Wars, as both Britain and France tried to break the American government's policy of neutrality, residents often found themselves facing the prospect of war with one or the other belligerent. In 1794, when war with Britain seemed likely, throngs of anxious citizens headed for Governors Island to erect fortifications (the old colonial-era fort on the Battery having been removed several years earlier); the same thing happened in 1798, when everyone expected war with France. After 1800, when Britain was again the probable enemy, work began on a string of forts that were expected to protect the city from naval attack for the foreseeable future: the circular West Battery (later Castle Clinton, and still later Castle Garden, originally built on a rocky outcropping about 200 feet off the Battery but now enclosed by landfill); Castle Williams and Fort Jay on Governors Island; Fort Wood on Bedloe's (Liberty) Island; the North Battery, which stood on the Hudson shore at Hubert Street; and further up river, Fort Gansevoort on Gansevoort Street. Ironically, when the United States and Great Britain did go to war in 1812, the city would be thrown into turmoil by reliable reports that the enemy intended to attack overland from Canada, not by sea. For a month or so in the summer of 1814, thousands of panicky New Yorkers turned out to build new forts, breastworks, and blockhouses on Brooklyn Heights, Upper Manhattan, and other locations around town. The invasion never materialized, but ever since the first city wall went up in 1653, residents had learned not to take such dangers lightly.

New dangers emerged during the Civil War, when the city's close ties to the cotton-producing South made it a veritable battleground between Union and Confederate sympathizers. In July 1863, thousands of heavily armed federal troops, many fresh from the recent fighting at Gettysburg, were brought in to suppress the Draft Riot—a four-day upheaval marked by savage street fighting, lynchings, and unprecedented destruction of property. Then, in November of the following year, rumors of a Confederate plot to burn the city prompted authorities to deploy additional thousands of soldiers at key points around town—to no effect, as it happened, for once the soldiers pulled out, Confederate arsonists set fire to hotels, theaters, stores, factories, and lumberyards, triggering a short-lived panic.

By the middle of the nineteenth century, however, two developments were already conspiring to obscure Lower Manhattan's long history of violent encounters with the outside world. For one, steady improvements in military and naval technology required New York's defenses to be situated further and further away from its shores—down to the Narrows, then out to Sandy Hook, Navesink, and the Rockaways, out of sight and mind. By the end of World War II, the last of the big guns protecting

Manhattan was removed altogether, testimony to the new reality of long-range air power and the intercontinental ballistic missile. Concurrently, the increasing density of trade and commerce in Lower Manhattan was erasing virtually every trace of its history as a residential community. As affluent denizens fled uptown in search of quieter, greener enclaves, their former homes became crowded boarding houses and tenements for the laboring poor, who were themselves soon displaced by the construction of new warehouses, shops, and offices. By 1850, only a handful of families still inhabited the blocks below Chambers Street—Manhattan's oldest neighborhoods, now inundated five days a week by commuters rooted in other neighborhoods, other towns, other states. (Surely one of the more telling facts about the thousands who died at Ground Zero was that relatively few lived anywhere near there.[2])

It is perhaps not so surprising, then, that the attack on the World Trade Center should have been interpreted as a decisive break with the past, the end of a long period of untroubled isolation from global events and a warning of new, almost certainly dangerous engagements ahead. Yet in trying to gauge the impact of September 11 on the future of New York, it may be helpful to remember that this is an illusion bred of our chronic present-mindedness—that, in fact, the city has always been deeply implicated in the world, and that this is not the first time its people have paid a heavy price as a result.

---

[2] These trends did not necessarily mean that New Yorkers stopped thinking of the entire city as a prime target for enemy attack, for in the decades just before and after the turn of the century, New York's destruction became a favorite subject of speculative fiction. Thus Arthur Vinton's *Looking Further Backward* (1890), a parody of Edward Bellamy's *Looking Backward* (1888), imagined a Chinese attack on Manhattan in the year 2023 that slaughters four million residents. Similarly, H. G. Wells's *The War in the Air* (1908), envisioned a German air fleet bombing the city into fiery oblivion. Although it is beyond the scope of this essay, I cannot resist the observation that these fantasies of municipal cataclysm continue to be recycled in such contemporary films as *Independence Day* (1996) and *Godzilla* (1998).

Whose Downtown?!?[1]

THE ATTACK ON SEPTEMBER 11 provoked New Yorkers' best qualities—a combination of toughness and generosity, street smarts, and worldly perspective. Friends, colleagues, families, and especially strangers suspended their daily routines to enact a larger sense of purpose. We had a moment of simple clarity amidst the daily clutter of artifice. What is important? What is not? Massive terror forced each of us to live a moment stripped bare of all illusions. And the suddenness of our loss created a need to reach out, to talk, and to pull together.

Downtown we became part of a creative, public outpouring that was especially tangible at Union Square. Quickly assembled materials—Santeria candles, color-copied photos of missing loved ones, flowers, poems, iconography displayed by lay Catholics and Buddhist monks, and artists' constructions—expressed the sentiments of tens of thousands. Acts of public goodwill and kindness flourished. Even the poorest gave something. Chinatown residents, many of whom cannot speak English and barely make enough for their own families, immediately raised $1.44 million for the relief fund. Green-card holders remarked how they felt "American" for the first time. This version of "we"—with millions of grievers and millions of consolers—was genuine and true.

Then the usual cleavages resurfaced. The Parks Department cleared away the memorials. At Ground Zero, the firemen were told to go home and scuffled with police charged

---

[1] For Vasu Varadhan and in memory of Moe Foner.

with keeping the site clear. A large cleanup company with personal ties to the former mayor was brought in to replace the volunteers. The postal workers at the Morgan Sorting and Distribution Facility in Midtown were ordered to go back to their anthrax-tainted workplaces. Anti-Arab and anti–"Arab-looking" backlash flared in the streets and online. While the police and FBI carried out secret roundups and interrogations, gas-guzzling SUVs sported American flags. And of course, the insurance compensation lawsuits began. Some victim's families contested the actuarial logic: lost mothers should not be paid less compensation than lost single men, nor should fast-food workers be considered less worthy than stockbrokers.

Anyone living and working downtown could feel the devastating and differential impact on those people, neighborhoods, and small businesses least able to survive the aftershocks. At five months and counting, many people still didn't have phones. While so much news and details about September 11 once saturated the airwaves, suddenly all but the most iconic stories were dropped. Incessant coverage gave way to back page clips. Those waves of ongoing grief were reduced to isolated pockets of mourning and remembrance. Official statistics can't begin to assess the long-term impact upon workers, immigrants, the poor, and the homeless. One even begins to wonder whether anyone cares. The damage was only hinted at by the growing unemployment statistics. The city, the state, and the federal government all declared the air quality safe, despite lingering foul and dangerous fumes. But amidst all this, the grinding, debilitating divisiveness of business—and life—as usual returned.

But wait: the spontaneous coming together at Union Square reminds us of the street performances that flourished in the seventies. Yes, this is what we love most about the city—the constant improvisation in people's intermingling and jangling. Out of such dissonance, new poetics cohere, becoming subcultures and movements. We can evoke the Beat poets, the Harlem Renaissance, or Walt Whitman, yet they are only contributors to a larger esprit, one that has made Manhattan so special. Even before the Dutch "discovery," this was a place of abundance and public exchange. Over the past several hundred years, as now, the port city drew people from everywhere. A vibrant, practical culture of juxtaposed differences emerged. It is this port city that has made Gotham so vibrant and

Community, shrine, memorial: the public space of Union Square Park, September 2001.
Photo by Richard Rosen.

inventive—the antithesis to suburbs and rural life. Yet despite the vi-
brancy, each period of history that has made and remade Lower Manhattan placed a priority on power and exclusion, and each gained
momentum from earlier inequalities.

Earlier historical struggles are usually understood as isolated historical
events, but they form a larger pattern. Together they compose a history of
the ongoing struggles over what values would shape the city and especially Lower Manhattan. The ruling logics of displacement and dispossession have won many battles, yet like a stubborn weed the port culture
resurfaces time and again. Let us begin with the World Trade Center and
then move back in time, examining some of those struggles long since
forgotten.

In the era that ended on September 11, New Yorkers who moved
through the WTC complex experienced the space as a hive. Most of us
entered from underneath, in a generic mall, chaotically bustling with
commuters and shoppers. The subway lines, the PATH trains, and the
chain stores served as the entryway. Tens of thousands of workers,
tourists, and visitors swarmed into the complex from below and moved
upward into and around the layers of worker cubicles above. This three-
decade youngster seemed so permanent—in part because it was built on
such an overwhelming scale. But it was the product of a relatively recent
dream of power, enacted wholly by the city's elites.

In 1958 the Rockefeller brothers proposed a billion-dollar redevelopment plan targeting 564 acres below Canal Street. A *New York Times* editorial affirmed the plan's modernist ambitions.

> Some of the poorest people live in conveniently located slums on
> high priced land. . . . A stone's throw from the stock exchange, the
> air is filled with the aroma of roasting coffee. . . . Such a situation
> outrages one's sense of order. Everything seems misplaced. One
> yearns to re-arrange the hodge-podge and put things where they
> belong.

Influenced by the orderly superblock designs of Le Corbusier, powerful
special interest groups sought to bulldoze those less profitable old build-

ings on irregular winding streets and replace them with large box super-
structures. Such plans were but the latest scheme to move residents, manufacturing, and small businesses out of the way. Robert Moses built the
highways, bridges, and tunnels to reorganize the region for these same development interests. In the name of the American dream, the white, up-
wardly striving classes were lured out to newer and newer suburbs. Today,
we take this suburban life for granted, assuming it flowed from the collective initiative of millions. We easily forget that such massive social engineering was decided from on high.

In the 1940s and '50s, downtown "renewal" schemes ripped out the
historical heart of Manhattan—the early, low-rise, mixed-used port-
culture buildings. Like the architects of a conquering empire, the planners of the WTC built upon a previously beloved place, a vast public
gathering place called the Washington Market, which stretched over the
Lower West Side. Built in 1812, the Washington Market was visited by
generation after generation of New Yorkers and visitors alike.

Public markets and specialty trade districts once flourished throughout
Lower Manhattan. The Washington Market was easily accessible from the
streets, creating a people's commons. In 1862, butcher Thomas De Voe
documented the early decades of these largely forgotten gathering places.
De Voe chose this 1814 verse to express his feelings about such markets:

> The place where no distinctions are,
> All sects and colors mingle there,
> Long folks and short, black folks and gray
> With common bawds, and folks that pray,
> Rich folks and poor, both old and young,
> And good and bad, and weak, and strong, . . .
> The high, the low, the proud, the meek. . . .[2]

Such a public space promoted a generous urban spirit of access and general well-being.

---

[2] Thomas De Voe, *The Market Assistant*, 7–8, New-York Historical Society Library.

From the eighteenth through the mid-twentieth century, this intermingling of diverse peoples and cultures characterized the port city. This
culture flourished in taverns, such settlements as the Five Points, the public markets, the parks, the Lower East Side, the streets, and the subways.
The industries of the port and scores of manufactures provided jobs for
newcomers and old timers. Conflicts abounded, but they were of a human scale. Going out into the neighborhoods was all about boundary
crossings. Dance halls, community organizations, union halls, and parks
were the venues to meet fellow countrymen and new neighbors. Where
else would Jimmy Cagney pick up phrases of Yiddish and Cantonese?
And you didn't have to be wealthy to survive and live well in Lower Manhattan. You didn't have to attend college to be educated about the world.

But in 1956, the Washington Market was shut down. Its wholesale
operations soon moved to the remote Hunts Point in the Bronx. The finance, real estate, and insurance interests prevailed, and Downtown became theirs to shape. In this sense, the destruction of Washington Market
was but a recent assault on the city's historical port culture.

Some of this hodgepodge was commercial; most was ethnic. The
WTC most immediately displaced Radio Row, which housed small manufacturers and specialty stores where hobbyists and engineers bought and
sold electronic parts. This 1950s displacement was preceded by the earlier
displacement of the city's Syrian community, which first arrived in the
1870s. Christians and Moslems from such places as Mount Lebanon,
Tripoli, Aleppo, and Damascus settled around Rector and Washington
streets, selling goods as part of the extended Washington Market. By
1924, it was a community hub for some 15,000 people. Robert Moses
wanted to build a mammoth bay bridge from Lower Manhattan to Governor's Island to Brooklyn. This required destroying Battery Park, the sacred historic settlement of Dutch and later British colonists, along with
sections of the Lower West and East Sides. The descendants of these colonial elites fought his designs. The extant fort, Castle Garden, and the
green were a direct link to the founding of Nieu Amsterdam and New
York. How could those landmarks be destroyed? After a bitter fight, the
decendants won. The Brooklyn-Battery Tunnel was built instead. The
residents of Little Syria, however, lost with either plan. Their protests

were to no avail. Expediting auto traffic to suburbia took priority over the community of politically powerless city dwellers. The Syrians had to leave and restart their businesses and their lives somewhere else. Many moved across the harbor to Atlantic Avenue in Brooklyn.

But other logics contributed to the evisceration of port-city culture. The 1924 Immigration Restriction Act championed by Protestant fundamentalists shut off the immigration supporting the then-thriving port zone of Lower Manhattan. The twenties were an era of 100 percent Americanism, a xenophobic movement of cultural cleansing driven by eugenic obsessions: Keep out undesirables. Prevent "race mixing." Drop old-country languages and customs. Promote absolute acculturation to WASP norms. Jews, Eastern Europeans, Italians, Greeks, Middle Easterners, Asians, Africans, Caribbeans, and Latin Americans were all targeted. Northern and Western European immigrants were given preference. The immigrant communities of Lower Manhattan—Syrian, Jewish, Italian, and Greek—faced a stark choice: lose one's roots and become modern, assimilated, and "white," or become part of a redlined inner-city enclave.

And even this bout of hyperinsularity paled in comparison to what had happened just a few decades before to Chinese New Yorkers, when burgeoning anti-Chinese racism culminated in the 1882 Chinese Exclusion Act. After a substantial Chinese immigration wave after the Civil War, Chinese laborers were suddenly deemed incapable of "self-ownership" and not fit for entry or citizenship. The act was not repealed until 1943, when China fought as a U.S. ally against Japanese imperial aggression. However, a quota of 105 persons per year of Chinese ancestry (from whatever national citizenship) was imposed. This restriction was actually not changed until 1968. Eighty-six years of exclusion choked off a thriving downtown settlement of Chinese mariners, cigar wrappers, merchants, and laundry workers, many of whom had intermarried with Irish women. A dispersed community became ghettoized in Lower Manhattan, in the Chinatown we take for granted today. Before Chinatown became a tourist site, it was demonized as a shadowy, evil underworld antithetical to Western civilization. The earliest signs of the port culture's gradual strangling, then, are apparent as far back as the nineteenth century.

But the power of downtown redevelopment was always premised on

moving the poor from high-priced land to ghettos. From the city's incep-
tion, European men of property could pursue life, liberty, and the ownership of others. Indeed, the founding of the city was premised on the
disenfranchisement of American Indians, blacks, and women. The Dutch
and the British established basic exchange systems that shaped subsequent social and spatial relationships. Property, the market, and subjuga-
tion determined one's rights and one's spatial settlement—logics that
continue to rule Lower Manhattan.

Downtown blacks were both dispossessed and displaced in the mid-
nineteenth century, just as the Chinese were after them. Frederick Dou-
glass described an 1840s New York where the original black community
still lived and worked around present-day Chinatown and westward into
the Five Points district, where the civic buildings now stand. And after
the 1863 Draft Riots, blacks were displaced further and further up the
West Side to Harlem.

This dispossession was hardly new to the city's black population. As
far back as the 1630s, African male slaves in Nieu Amsterdam maintained
the fort and wall, cleared land, brought in harvests, and performed other
labors. African women worked as domestics. Unlike British slaves, however, Dutch slaves could own land, testify in court, bear arms in times of
emergencies, and marry in the Dutch Reformed Church. And while Governor Kieft ruled they could not intermingle with Europeans, he did
grant them "half freedom" in 1644. Some gained deeds to *bouweries*, or
livestock farms, making annual payments in foodstuffs, yet their children
remained the property of the Dutch West Indies Company. One farm
was on a former Lenape Indian encampment by the Kalch-hook just west
of present-day Chinatown. Such farms supplied the downtown meat
markets and served as buffers against angered up-island Indians.

Under subsequent British rule, prior slave rights (and the rights of free
women) were eliminated and absolute ownership rules codified. By 1730,
enslaved Africans and Afro-Caribbeans became a major trade commodity
in New York. In a rapidly growing city serving a growing regional marketplace, labor was always in demand, and black women served as manual
and sexual chattel. Individual businessmen such as Knickerbocker Freder-

ick Philipse, the most prominent slave merchant in the city, candidly explained: "It is by negroes that I find my cheivest [chiefest] Proffitt. All
other trade I only look upon as by and by."[3] By 1741, one in five of the
11,000 New Yorkers was a slave. Profit, power, and one's social identification elevated some at the cost of debasing various others.

We now arrive at Manhattan's foundational moment of violence, displacement, and dispossession. Dutch Governor Minuit's fabled $24 pur-
chase from unnamed American Indians has been repeated, as with all
good stories, *ad nauseum* until it has become part of a mythic collective
identity. Even when debunked, this is "our" commercial creation myth:
the shrewd Western merchant triumphs over the unknowing primitives.
Thereafter, rapid material "progress" culminates in the building of the
World Trade Center—the ultimate symbol of Gotham's rise to global
power. Yet a more accurate story must be told.

The Munsee Lenape viewed this land as held in common trust for the
creator Kishelemulong, and it could be neither sold nor owned. Hunting
and planting lands could be shared or temporarily transferred. Exchanges
in gifts did not signify purchase but reciprocal obligations—human, spir-
itual, and ecological. In the first years after the Dutch arrival, cultures co-
existed and cross-mixed. Dutch moralists warned against such
interminglings, but this daily interaction formed the basis for peaceful
cosurvival.

Differences, however, soon came to dominate in Lower Manhattan.
Ecocultural conflicts escalated. The woodland habitats for hunted ani-
mals disappeared as more Dutch West India Company settlers cleared
farmlands and erected fences delineating their private boundaries.
Colonist cows and pigs trampled native fields, and native domestic dog-
wolves attacked grazing livestock. New diseases decimated the Lenape to
one-tenth of their original population. Their hunting and planting culture's rapid decline forced on them a greater dependence on wampum
making. Within two generations, the Munsee Lenape went from a ro-

---

[3] Cited in Joyce Goodfriend, *Before the Melting Pot: Society and Culture in Colonial
New York City, 1664–1730.*

bust, independent people aiding struggling Dutch West Indies traders to facing a new law-and-order governor who demanded tribute for "protecting" them. Living off nature's local bounty had been transformed into an international profit transaction.

Governor Willem Kieft, much reviled in his day and by historians thereafter, should properly be credited as a patriarch of modern, capitalist New York City. From his arrival in 1638 to his replacement by Stuyvesant in 1647, Kieft performed the Protestant disciplinary work of Dutch colonization. He carried out the company's decision to "buy" Lenape land and authorized the first private land ownership by "free people."

From 1640 to 1645, Kieft launched a series of attacks on local Indians that provoked Indian retaliation, which in turn sparked even more brutal company massacres. On a February night in 1643, company men descended upon two camps of sleeping Munsees, one by Corlear's Hook and one by present-day Jersey City. A Dutch Staten Islander, David de Vries, gave witness to the slaughter:

> Infants were torn from their mother's breasts, and hacked to pieces
> in the presence of the parents, and the pieces thrown into the fire
> and in the water, and other sucklings, being bound to small [cradle] boards, were cut, struck, and pierced, and miserably massacred
> in a manner to move a heart of stone. Some came to our people in
> the country with their hands, some of their legs cut off, and some
> holding their entrails in their arms. . . .[4]

Such was the originating cost of this new private-property system that free-marketeers so romantically celebrate today. Kieft's War effectively cleared Manhattan's land for future European settlement, and between 1640 and 1664, that population increased eighteenfold. These are the true origins of modern New York City.

Certainly, none of the power hierarchies emerged from simple conspiracies. Instead, they grew out of the regular, systemic exercise of accu-

---

[4] Franklin Jameson, *Narratives of New Netherland, 1609–1664* (Charles Scribner's Sons, 1909), 227–29.

mulated power and decision-making. Advantages of control and profit governed this calculus of power. Those most different from the norm of individualist material progress have had it worst. It was these norms that created and regulated the realities of what got built, whose interests were protected, and whose were disenfranchised. And these basic principles, although much modified over the centuries, have remained ascendant in Lower Manhattan.

Downtown, business has been returning to normal. 9/11 has had a devastating impact, but another, far quieter displacement of massive proportions was already under way. The dot-com frenzy and gentrification pushed out lots of garment factories around Chinatown as well as long-time residents of the Lower East Side. In the name of "undervalued square footage," loft by loft, apartment by apartment, workers and residents have been displaced. Many of these "upgraded" spaces are now empty, waiting to be rerented at a premium. Landlords would rather sit on property and hope for a new boom than run the risk of renting to poor people who might prove difficult to evict. If the ruling logics of finance, insurance, and real estate continue to go unchallenged, the remaining port-culture communities of old New York will be permanently wiped out.

We can't just remember those who perished and have been displaced on 9/11. We need to learn from past injustice. Grand top-down schemes can be challenged. Jane Jacobs's fights for a "healthy city" protected the West Village from bulldozers. Greenwich Villagers, Italians, and Chinese residents fought Moses's Lower Manhattan Expressway. We take the walkable, human-scale, mixed-use remains of Downtown for granted, yet they would have been torn down if those battles had not been fought. We would not be enjoying access to the Hudson River now if it had not been for those long years of fighting Westway.

The intermingled port cultures of Lower Manhattan have yet to be understood as a living historical cultural resource, let alone protected. Today, Chinatown and the old Lower East Side remain as surviving links to this core New York City culture—communities struggling to hang on while the forces of gentrification and displacement beckon. It is here where the democratic, public spirit of intermixing still flourishes. Now is the right time to give this core New York port culture substantial recognition and support. All of it should become a historic district. All of it should be protected from gentrification. The historic preservation of neighborhoods must enable regular people to continue living there. We don't need shopping malls in historic buildings. We don't need more big box office buildings that further segregate work from everyday life. Fair, human-scale development can be ours. Together with prudent historical landmarking strategies and community-based activism, disposession can be halted in its tracks.

The Port Authority, the city, the state, and the private sector have to be held accountable to a higher standard—a standard guided by principles of sustainable development, social justice, and an honest history. If we can harness that amazing collective goodwill we created immediately after the 9/11 attack, and give voice to those who have been displaced, we can certainly rebuild a Downtown guided by human-scale values truly working for an inclusive public good. We need a development logic that cultivates community and the human spirit. Remember Union Square!

The First Wall Street Bomb

A FEW BLOCKS EAST and a few blocks south of what was once the World Trade Center, there is a memorial to the victims of terrorism. It is easy to miss—just a dozen or so pockmarks the north side of the old Morgan Bank. There is no sign, no plaque, no list of names. For the past 81 years, the pitted marble wall at the corner of Wall and Broad streets has served as the sole public marker of what was, until September 11, 2001, New York City's worst terrorist disaster.

The attack, known as the Wall Street explosion, took place on September 16, 1920. Like the recent assault on the World Trade Center, the 1920 explosion struck the heart of the city's financial district. It killed more people than most Americans imagined possible from a single assault. It seemed to leave the nation facing an unknown enemy accused of committing an act of war. And, like the destruction of the World Trade Center, it gave New Yorkers and Americans a queasy sense that their world had changed irrevocably.

In 1920, the corner of Wall and Broad streets marked the center of American finance capitalism. On the northeast lot stood the Greek Revival building of the government Sub-Treasury, now the Federal Hall National Memorial. Across the street, on the northwest corner, the offices of Banker's Trust soared 35 stories from base to roof. On the southwest lot stood the New York Stock Exchange, with its feminine frieze of Integrity sheltering the nation's miners and farmers. A small Italian Renaissance building on the southeast corner housed the Morgan Bank, the seat of the country's financial power.

Just after noon on September 16, as hundreds of workers poured onto the street for their lunchtime break, a horse-drawn cart exploded in a spray of metal and fire. "It was a crash out of a blue sky," wrote one witness, "an unexpected, death-dealing bolt which in a twinkling turned into a shamble the busiest corner of America's financial centre." Survivors remembered a flash, a roar, then silence. Wall Street seemed to freeze in a tableau vivant. Runners lay flattened like tenpins. Curb brokers stopped mid-transaction. Stocks and bonds hung suspended in the haze. Then windows for blocks around burst from their frames, launching a panicked run for cover.

Veterans likened the scene to a battlefield during the Great War. Men and women ran and crawled from the bomb site in widening circles, north on Nassau, west on Broadway, east toward the river, and south on Broad. Some victims said they didn't feel as if they were trying to move at all, more that fright and shock and the force of the crowd had borne them along. On the Wall Street side of the Morgan Bank, an overturned touring car snapped and sizzled beneath a tower of flame. The fumes from the car lent a sharp chemical tang to the odors of blood and burning flesh that spread through the neighborhood. On the steps of the Sub-Treasury, a bronze statue of George Washington gazed down on the carnage like a paralyzed traffic cop, hand raised halfway in a futile plea for order.

Among the wounded were teenage sisters Margaret and Charity Bishop. Both suffered severe burns while on their way to lunch. Margaret was expected to die. Charity would live, the doctors said, but would "doubtless carry the marks of her injuries for some years." Charles Lindrothe, a newly married clerk for the National City Bank, had survived gassing in France during the war, only to meet his death on Wall Street. William Hutchinson, an insurance broker from Garden City, Long Island, left behind a wife and three-year-old daughter. William Joyce, a young Morgan employee, had died instantly, pinned to a cage on the main banking floor like a butterfly captured for display.

Such fragments of information, revealed in newspapers' victim lists and articles, revealed a universe of grief, a disruption of shared expectations. "The girls just went to work," cried one mother, "and here they are at the point of death." By the day after the blast, approximately thirty people were dead, and by the end of the year forty had lost their lives. The attack, wrote the *Sun* and *New York Herald*, was "unprecedented in horror."

That the numbers of dead may now seem paltry—statistics from a past when we counted death in dozens instead of thousands—underscores just how violently our own expectations were disrupted on September 11. Today, the destruction of the World Trade Center stands alone in the annals of American terrorism. The immediate horror has led many people to denounce it as the purest evil, a paroxysm of hatred without context or precedent. We say that destruction on such a scale, arriving in the course of an ordinary day, is unthinkable, unfathomable, unimaginable. But the questions that it poses, the responses that it has so far provoked, and even the language we use to describe it, are not themselves without precedent. Horror, too, has a history.

When it comes to mining the past for lessons in the present, the differences between September 1920 and September 2001 are both insurmountable and insignificant. The United States today is a far more powerful global force, economically and militarily, than it was in 1920. The character of the so-called enemy within has shifted, and whatever threat left-wing radicals once seemed to pose has been replaced by an assault from the right. One would like to think, too, that contemporary Americans are more sensitive than our predecessors were to issues of racial and ethnic repression, less inclined to resort to scapegoating or internment camps.

But if the political context has changed since 1920, the imperative to understand violence within a political context has not. However random or evil an act of terrorism may seem, we can rest assured that the men who commit it do so for certain reasons, and that our response will be conditioned by our own political culture. Our current obsessions—with civil liberties, intelligence, nationalism, discrimination—did not spring fully formed from the ashes of the World Trade Center. What seems so new about this historical moment—our sudden fear of a shadowy foreign enemy working within U.S. borders—is also very old.

In 1920, Americans believed that they, too, faced a serious and violent threat to their national institutions. Their response to that threat will not, undoubtedly, be our own. But ours, like theirs, will be crafted at the in-

tersection of necessity and ideology. One advantage of history is that it allows us to sort out, from a distance, which is which.

For all of our professed ignorance of our vulnerability, of the depth of
anger in the Middle East, of the technology and politics that made the
World Trade Center disaster possible, there are hundreds of hints that,
before September 11, we were not quite as oblivious as we would now
like to believe. Feature films played out scenarios in which cities fell to
terrorist siege. News reports described warfare that would be fought
through stealth attack. September 11 wasn't even the first time that terrorists had hit the World Trade Center. The current crisis may feel unexpected. In retrospect, though, it appears we anticipated . . . something.

Americans expressed a similar contradiction in 1920. While the Wall
Street explosion seemed unprecedented, it also seemed utterly pre-
dictable. Everyone had heard anarchists and communists raving about the
depradations of Wall Street. Private detectives and Department of Justice
agents had warned for months of an impending attack. The previous year,
bombs had exploded almost simultaneously in seven American cities,
hinting of a frightening new conspiracy. At the sites, the bombers left behind pink leaflets, titled *Plain Words*: "There will have to be bloodshed;
we will not dodge; there will have to be murder; we will kill, because it is
necessary; there will have to be destruction; we will destroy to rid the
world of your tyrannical institutions." It was hard to say that Americans
hadn't been warned.

One man had even predicted the Wall Street explosion itself, almost to
the day. Edwin Fischer, a former New York City tennis champion and
lawyer, had sent postcards to several friends on Wall Street, warning them
to "get out" and "keep away" on the afternoon of September 15. Police
arrested him soon after the explosion, but within days they had declared
him innocent. They said his prophecy was just a coincidence. In his "eccentric" mind, Fischer had simply transformed a widespread fear into a
weirdly specific prediction.

But while Americans in 1920 expected some kind of attack, they
claimed afterward that they had never imagined violence of the sort that
erupted on Wall Street. Earlier bomb explosions—the Haymarket in
1886, the *L.A. Times* dynamiting in 1910—had been part of specific la-

bor disputes, party to specific protests. The blast on Wall Street, by contrast, appeared to be purely symbolic, designed to kill as many innocent people as possible in an assault on American power. "There was no objective except general terrorism," wrote the *St. Louis Post-Dispatch*. "The bomb was not directed against any particular person or property. It was directed against the public, anyone who happened to be near or any property in the neighborhood." The men and women killed were not captains of finance. They were clerks, stenographers, brokers, messenger boys. J. P. Morgan, the figurehead of American finance capitalism, wasn't even in the country on the day the blast occurred.

The *New York Tribune* wrote that a "new anarchism" had come into the world. The *Washington Post* called it an "act of war."

But if this was war, it was, like today's violence, war without a definite enemy. Most Americans thought they knew the enemy in some general way: the bomb had been set by the very people who had long threatened to do so. Today, that logic yields Osama bin Laden, who has repeatedly assured the world of his violent hatred for the United States. In 1920, suspicions focused on anarchists, communists, and various other shades of Red.

Given the lack of hard evidence uncovered in the Wall Street rubble, the suspicion that radicals had been involved said as much about contemporary fears and politics as about the motives of the supposed bombers. As an unsolved mystery, the explosion was a Rorschach test of sorts—a series of images and events onto which American could, and did, project any variety of anxieties. It was also an event with great political potency, an opportunity for blame to be cast, money to be disbursed, power to be reapportioned. These two aspects were intimately linked: how Americans defined the threat helped to determine, ultimately, how they responded to it.

In the days immediately after the Wall Street blast, most newspapers called for reason and calm, arguing that the best response to terror was a demonstration of liberty's virtues. The blast inspired calls for national unity, for a show of democratic resilience. If the country could not carry on as before, it was said, the terrorists would have accomplished their aim. "Whatever may be discovered to be the nature of the catastrophe," the *New York Post* declared, "it is plain what the proper attitude of the public mind should be at the present moment. And that is a firm resolve to go on with the business of the city and of the country while waiting for the facts upon which the law must proceed."

In 1920, such appeals were by no means mere statements of principle. The men who wrote those lines had a clear and present example of what happened when Americans did not simply "go on with business." Following the nationwide bomb plot of 1919, Attorney General A. Mitchell Palmer had geared up for a major assault on suspected immigrant radicals. One of his first steps had been to place 24-year-old J. Edgar Hoover in charge of a new "radical division" dedicated, with ample funds, to ridding the country of subversive elements. The high point of the backlash had come in November 1919 and January 1920 when federal agents, local police, private detectives, and volunteers swooped down on radical gatherings across the country, netting thousands of suspected Reds.

The Palmer Raids, as they came to be known, focused on noncitizen immigrants, who could be arrested and deported without the bother of a due-process trial. With great fanfare, undesirables such as Emma Goldman and Alexander Berkman, who had lived in New York for years, were shipped out of the country. By September 1920, though, it was evident that the raids had resulted in extreme civil liberties violations—without, apparently, making much of a dent in the bombing problem. Palmer vowed to do better on Wall Street.

The first order of business was to determine who, within the general Red category, was personally responsible for the blast. The Bureau of Investigation, the forerunner of today's FBI, quickly declared that the explosion had been the work of the Galleanists, a group of Italian anarchists also suspected of setting off the 1919 bombs. The New York police, by contrast, hesitated to commit to a culprit, suggesting at first that the explosion might have been an accident. Eventually, they acceded to the idea that radicals were appropriately suspect. William J. Burns, a famous private detective, announced that Russian communists were responsible. Newspapers extended blame to include "the Bolshevist-minded, . . . both parlor and non-parlor varieties," and assorted other "propagandist foe[s] of our institutions."

Within these various suppositions, though, all agreed that the bombers were likely possessed of certain essential qualities: they were for-eign-born, or at least foreign-influenced. They cherished delusions about the overweening power of Wall Street and the evils of American capital-ism. And they were, somehow, not in control of their own faculties. "Per-haps they had filled their minds with stories that Morgan & Co. made war, had magnified in their imaginations the importance of one little man," speculated the *Milwaukee Journal.* "They lack the intelligence to realize that these stories were only a plot of other men to gain their own ends by deceiving the ignorant."

This view of the bombers as deluded zealots, led astray by an irrespon-sible leader, served a variety of political purposes in 1920. It suggested that while political protest might have been the excuse for the blast, the violence was not, per se, political. Therefore, there was no reason to begin to address any of the political grievances one might imagine the bombers were attempting to express. The explosion "killed an uncertain number of absolutely innocent and uninterested people . . . ," wrote the *Wall Street Journal.* "It sent a sufficient number of injured to the neighboring hospi-tals to make a total record of casualties about equal to a raid in France in No Man's Land during the late war. And that is all it did." In 1920, as to-day, the impulse was to reject any analysis of "root causes"—to dismiss, defiantly and patriotically, calls for national self-examination.

Within this framework, the bombing served to affirm the virtues of the United States—its freedom, its toleration, its energy, its wide distrib-ution of property. "The average citizen is well off and he knows it," snapped the *Journal of Commerce and Commercial Bulletin* on September 18, "he recognizes perfectly well that he is . . . perhaps better off than any other equal body of people in the world." The bombers became a foil for all that Americans would like to imagine they were not. Reds were au-tocratic; Americans, democratic. Reds were violent; Americans, peace-loving. Reds were easily deceived; Americans were independent thinkers.

This view of the likely bombers as radical dupes established a political equation particularly generous to those already in power. National leaders such as Palmer sought to use the fear of violence to consolidate support for a renewed campaign against the Reds—a campaign for which, not in-

cidentally, the Justice Department would need a great deal of money and
leeway. Palmer's backers suggested that anyone who objected to such a
campaign was in effect siding with the men who had committed murder
on Wall Street. "The New York disaster teaches a powerful lesson," wrote
the *Washington Post*. "So long as the radicals, either of the parlor or sewer
variety, are permitted to preach theories that are inimical to the American
government, to foment class hatreds and inflame the minds of irresponsi-
ble people, just so long will the danger of such calamities exist."

Since the World Trade Center attacks, Attorney General John Ashcroft
has revived such arguments to build support for his campaign against immigrants suspected of untoward terrorist connections. Critics who
frighten the American public with the specter of lost civil liberties, he
told the Senate recently, "only aid the terrorists, for they erode our national policy and diminish our resolve." Ashcroft is depending for support upon the alleged surge in patriotism that has so far been the most
celebrated effect of the World Trade Center attacks. News reports hyping
the rise in patriotism may yet prove to be more wishful thinking than
tangible reality. Even so, the apparent increase in nationalism is far more
complicated and far more menacing than devotees of "greatest-genera-
tion" unity would have us believe.

Already, Ashcroft has used the World Trade Center attacks to justify
secret military tribunals, wiretapping of attorney-client phone calls, vast
expansions of the FBI's surveillance power, and the detainment of thousands of immigrants who have yet to be charged with any crime. In his
targeting of noncitizen immigrants, Ashcroft seems to be taking a page
from Palmer's book, using lack of naturalization as an excuse to deny war-
rants, due process, legal counsel, and public trials. He would do well,
however, to look elsewhere for historical role models. Palmer's popularity
lasted for barely six months after his infamous raids.

By September 1920, when the bomb went off on Wall Street, Palmer
was roundly regarded, in the words of one newspaper, as a man prone to
"cry wolf" on the subject of Reds. While he used the Wall Street explosion to appeal for a federal sedition law and a revival of the raids, Palmer

had spent his political capital on the earlier campaign, and even the Wall Street explosion was not enough to replenish the funds.

It might have helped if his Bureau of Investigation had been able to solve the crime, but despite dozens of arrests, they never uncovered the Wall Street bombers. Many radicals took this as evidence that the blast had been a plant by industrial interests to discredit the Left, or perhaps an accident being covered up by some dynamite concern. More likely, it was due to the sheer incompetence and disorganization of the federal Bureau, aided by a penchant to assign collective rather than individual blame to Reds.

Show trials and government campaigns are not always necessary, however, to turn celebrations of national unity into a serious threat to difference and dissent. If the Wall Street explosion did not produce a revival of the Palmer Raids, it did have serious consequences for American democracy. Nativists spun the explosion as yet another reason to support immigration restriction, especially of those groups—Italians, Russians, Jews—suspected of radical tendencies. "The bomb outrage in New York emphasizes the extent to which the alien scum from the cesspools and sewers of the Old World has polluted the clear spring of American democracy," wrote the *Washington Post*. The equation of violence and foreignness, and of foreignness and radicalism, helped to justify the discriminatory immigration laws enacted over the next several years.

Wall Street, too, ultimately capitalized on the blast. Coming at the end of the Progressive Era, when Wall Street's power was the subject of fierce political contest, the explosion helped to affirm the idea that loyalty to America meant loyalty to capitalism as well. To sympathize with the victims of the bombing, in other words, was to sympathize with Wall Street itself. Critics of the country's economic system were denounced for supporting violence and terror. "As a deliberate attempt at massacre," wrote the *New York Times*, ". . . it shows us what mad passions are ready to be stirred by the wild and whirling words of editors and politicians." Debate over Wall Street's power grew noticeably muted in the weeks after the bombing. It was in such silences rather than in shouting that the explosion had one of its most profound effects.

It would be absurd, in our current situation, to blame left-wing critics of capitalism for the actions of far-right religious zealots. But it is also impossible to imagine that for the foreseeable future, protest movements in this country will not be tainted with un-Americanism. Already, cheers for war have been accompanied by efforts to silence advocates of peace. Even the desire to return to the imagined past of "normal life," the defiant resolution to carry on as before, has something of a darker side. Affirmation of the status quo is, by definition, a form of resistance to change. Our determination to restore a measure of normalcy to our lives in the wake of the attacks can easily become willful blindness to the injustices, large and small, of daily existence.

Current appeals to "patriots" to keep their money in the stock market, to spend those consumer dollars, sound oddly similar to the calls, in 1920, to proceed with "business as usual."

After the 1920 explosion, the financial district, fearing a market collapse, determined to restore operations as soon as possible. Cleanup crews worked through the night of September 16 under searchlights, collecting debris, righting furniture, scrubbing away stains of blood. The sidewalks were swept clean of even the usual grime. By the morning after the blast, only an occasional lump of plaster, tumbling from a top floor, marred the eloquently sanitized scene.

The Morgan Bank and the Stock Exchange managed, remarkably, to open at the customary time on the day after the explosion. And thanks in part to support from bankers, who worried that the blast would spur financial panic, the stock market continued the previous day's rapid climb. One broker suggested that the bombing would actually improve Wall Street's fortunes: "Public opinion is now thoroughly aroused against radicals, criminals, and everything which is opposed to orderly government. Is not this bullish and favorable to industry, investments and the abolishment of radicalism?"

His predictions seemed to bear themselves out, at least in the short term. By September 20, the *New York Commercial* could report that: "There is unmistakably a better feeling in Wall Street than there has been for some time."

After the World Trade Center attack, the market—as well as the rest of the U.S. economy—did not fare quite so well. But when the men and women of Wall Street returned to work on September 17, 2001, nearly a week after their neighborhood exploded in fire and debris, they unwittingly reenacted a pattern set many years before. September 17 was the same day that Wall Street had gone back to business after its first terrorist disaster, 81 years before. In the opening bell of the Stock Exchange, we heard an echo of history.

The 1920 explosion failed to divert or even much disrupt New York's rise from a provincial center of capital to a global financial power. It did suggest, though, that this concentration of power would make Lower Manhattan a target of political violence.

New York had to learn that lesson anew on September 11. For thirty years, the Twin Towers of the World Trade Center stood as cocksure symbols of Lower Manhattan's global economic power. When they fell, they took with them the amnesiac's confidence that history, like skyscrapers, moves only in one direction, ever upward to new heights and glories.

This new sense of vulnerability does not mean that New Yorkers, for whom the attack has been personal as well as national, must turn inward like their predecessors in 1920, rejecting "foreign" people and "foreign" ideas. Repression and racism, certainly, have a long and less-than-hallowed history in the city. But New York has an equally strong tradition of willful diversity, a collective faith that the world's peoples can crowd into a few square miles and still get along.

The destruction of the World Trade Center has forced New Yorkers to choose, consciously, between these two traditions. The 1920 explosion helped to usher in an era of fierce reaction and conservatism, in which immigrants, radicals, and reformers were pushed to the city's margins. As today's New Yorkers begin to reimagine Lower Manhattan, craft new security policies, and cope with new suspicions, they have the opportunity to make a better choice.

Cracks in the Edifice of the Empire State

Desperate for news on the morning of September 11, I twiddled my radio dial and hit upon a station that was relaying accounts from the BBC. They reported events as an attack upon the main symbols of global U.S. financial and military power. This was such an obvious interpretation that I scarcely remarked upon it. But when I tuned in to the U.S. media, I noticed that no one took that line. In New York, understandably enough, September 11 was represented as a local disaster of horrific magnitude with unfathomable causes and unthinkably tragic personal and local implications. Nationally, the media immediately followed President Bush in construing it as an attack upon "freedom," "American values," and the "American way of life." An amazing consensus was quickly forged throughout the nation on that point to the exclusion of almost anything else. But when I switched back to the international media, the BBC's phrasing was far from aberrant or unusual. There were, clearly, different local, national, and international ways of understanding events.

A certain power resides in discourses. Once things are cast exclusively in one language, then it becomes difficult, if not impossible, to hear or see critical perspectives that depend upon another. To view September 11 in terms of personal tragedy or American values rather than as an attack upon global U.S. financial and military power obviously constrains both argument and understanding. Attempts to raise questions about the use and abuse of U.S. global financial and military power in New York teach-ins, for example, were quickly rebuffed in the New York media and by local officials with the suggestion that people should go tell that to the bereaved families of firefighters in Queens. Nationally, to raise such questions was construed as being contrary to freedom and the American way, and many instances of outright suppression of such thoughts occurred. I want to argue here, however, that the BBC perspective offers a useful counterpoint from which to probe into the nature of the events themselves while understanding some of the contradictions within the American response.

The Twin Towers symbolized the era of neoliberal globalization and the role of New York financial markets in particular and the U.S. in general in forcing a certain pattern of political-economic development (at one point known as "the Washington consensus") upon the rest of the world. They marked in towering glass and steel the moment of transition from Fordism to flexible accumulation led by financialization of everything.[1] They symbolized the newfound dominance of finance capital over nation state policies and politics. By the time of the Clinton presidency, it was clear that even the federal government had to submit to the discipline of New York bond markets.

In the early stages, however, the towers were mired in the problems of serious recession and stagflation, epitomized locally by the technical bankruptcy of New York City in 1975. The recovery of New York depended crucially upon the further deindustrialization of the city's economy and its single-minded devotion to and ultimate dependency upon financial services and ancillary activities (legal services, information processing, the media). Financial speculation and the invention of new instruments of debt and credit became the lifeblood of a city that was proudly global at one level but also increasingly parasitic upon the production of real values in the miserable workshops of Bangladesh, the *maquillas* along the Mexican border, and the tyrannical factories of South Korea, Indonesia, China, and Vietnam. New York's economy rode the neoliberal tiger in the 1990s. It became intoxicated with the "irrational exuberance" of financial markets. It extracted much wealth from the rest of the world to become the richest and the most spectacularly successful

---

[1] See David Harvey, *The Conditions of Postmodernity* (Blackwell, 1989).

of the global cities that came to operate as command and control centers for global financial flows. It became the center of economic empire. The astounding wealth accumulated in New York City forged a basis for a remarkably rich cultural life and levels of conspicuous consumption beyond belief. It became a tourist mecca as people flocked to see the spectacle.

But New York also developed a vast low-wage economy to service its needs in restaurants, in hotels and apartment houses, in shops and workshops, in transportation, and in the upkeep of the city. Wave after wave of immigrants from everywhere crammed into the city to feed on the crumbs from the tables of the wealthy financiers, lawyers, and media moguls. This other New York was perpetually on the brink of total impoverishment, its slowly rising incomes often more than offset by spectacular rises in rents and other living costs in one of the most economically successful urban economies of all time.

The towers were not, therefore, neutral or innocent spaces in the global or even local scheme of things. This does not mean, of course, that anyone there deserved to die. Workers of all sorts (from janitors, cooks, and rescue workers to bond traders) were there. People need jobs, preferably well-paid ones, and for the most part they just do what their job specification says they should do. Markets hide consequences and social relations very effectively. Many of those in financial services were insulated from the global consequences of their actions (I know people who pulled out of the industry when they did figure out what it was doing to people and environments on the ground). Daily life in those towers appeared innocent in part because of the way the two New Yorks intermingled there, but also because of the disconnections between actions and consequences in the global economy.

Intense local emphasis upon loss and grief feeds on and promotes the idea of innocence. The brilliant short obituaries in the *New York Times*, coming day after day for weeks after the event, celebrated the lives and special qualities of those that died, making it impossible to raise a critical voice as to what role bond traders and others might have had in the creation and perpetuation of social inequality either locally or worldwide. The intensity of the local discourse, in short, occluded such social commentary.

If freedom is exclusively defined in terms of market freedoms, of course, then the space of the World Trade Center could be (and was) depicted as a "space of freedom" delivering market opportunities to everyone in the world in the face of (largely oriental) despotism and state-led (largely communist) terror. The attack upon the World Trade Center was an attack upon this kind of freedom. But freedom is one of the more complicated and paradoxical words in our political vocabulary. Freedom of the market can hide and preserve a whole raft of negative freedoms such as freedom to exploit labor, to deplete and degrade environments and destroy habitats in the cause of profits, to extract high rents, to lay people off overnight with no security, to exercise private property rights ruthlessly in the pursuit of individual advantage.

For those who have a more jaundiced view of what neoliberal globalization and market freedom have really been about in these last few years, the towers therefore symbolized something far more sinister. They represented the callous disregard of U.S. financial and commercial interests for global poverty and suffering; the militarism that backs authoritarian regimes wherever convenient (like the Mujahadeen and the Taliban in their early years); the insensitivity of U.S.-led globalization practices to local cultures, interests, and traditions; the disregard for environmental degradation and resource depletion (all those SUVs powered by Saudi oil generating greenhouse gases and now, in New York City, adorned with plastic U.S. flags made in China); irresponsibly selfish behavior with respect to a wide range of international issues such as global warming, AIDS, and labor rights; the use of international institutions such as the International Monetary Fund and the World Bank for partisan political purposes; the shallow and often hypocritical stances with respect to human rights and terrorism; and the fierce protection of the patent rights of the multinationals (a principle that the U.S. enforced with respect to the AIDS epidemic in Africa but then cynically overthrew when it needed Cipro drugs to combat the anthrax menace at home).

But to say any of this not only offends against the local discourse of personal pain and suffering of innocent victims, it also runs against the grain of interpretations based on an attack upon American values and the American way of life. Oddly, more than a whiff of a new McCarthyism was apparent in the wake of September 11 as freedom to comment in this vein was actively repressed.

The odd thing with respect to "American values," however, was that it took the attack to rediscover values that had gotten lost under the dominance of the dot-com boom, Wall Street greed, and Madison Avenue gloss. New Yorkers, faced with unspeakable tragedy, for the most part rallied around ideals of community, togetherness, solidarity, and altruism as opposed to beggar-thy-neighbor individualism. The outpouring of help, monetary as well as tangible (the lines to donate blood were incredible), from all sources was absolutely extraordinary. An abrasive and divisive mayor was transformed into a ministering angel of the streets. Giuliani became an icon of comfort and common sense for everyone. It became possible for a short period of time to talk about the collective good instead of individual interests. Unionized municipal employees became heroes. Government, which we had been told *ad nauseam* these last twenty years was all bad, was suddenly looked to as a source of comfort and good. Its failure to provide security and to fund public health adequately suddenly became a problem. Tax-cut Bush, a Texan with no love for New York or for Keynesian public financing, immediately promised $20 billion and whatever else it took to rebuild Manhattan.

Above all, we had three days of noncommercial television, as if the country was collectively ashamed of its terrible habits of mindless consumption. For many of us that was a wondrous experience. The talking heads of cable TV forgot sex scandals and other stupidities and tried to take the world seriously (even if they made heavy weather of understanding why the rest of the world might hate us so). We suddenly found ourselves in a world where different (and for many of us much more worthwhile) values were being articulated—values quite at odds with the American way of life as experienced before September 11.

Much of this was wrapped up, however, in a more troubling patriotism and jingoism as betokened by the media's insistence that it was a distinctively "American" set of values that had been attacked. American flags festooned the city, though it was not always clear with what motivation (immigrants used them to protect against harassment, some used them as signs of local solidarity, and others as a sign of patriotic fervor). And the words "In God We Trust" rivaled "God Bless America" (as if this was somehow qualitatively different from the idea that "Allah Is Great") as rallying points for civil solidarities. Dissident views were condemned outright, and civil liberties and freedoms of speech were threatened. There were more than a few signs of U.S. versions of fanaticism and zealotry, initially directed against Muslims. But Mayor Giuliani for one immediately condemned this: the last thing he needed was intercommunal violence in a multiethnic city.

But then the economy turned sour, locally in New York City as well as nationally and internationally. Calls to patriotism did not avert a market plunge when the Stock Exchange reopened, newfound solidarities did not prevent mass firings from the airline, travel, and "hospitality" industries, and close to a 100,000 jobs were lost in New York City alone in the wake of September 11. Financial services that had half a foot in New Jersey or Connecticut anyway completed their moves out, and surpluses of office space were identifiable in Midtown Manhattan even after the losses of Lower Manhattan had been compensated for. September 11 appeared more and more as a wonderful excuse for companies and industries to do what they were preparing to do anyway (including moving out of a highly congested and very much overpriced Manhattan).

Social inequality in the city became more rather than less emphatic. The proposed financial compensation scheme for those who died in the World Trade Center disaster illustrates the problem. Although much praised for its moderation, the government plan envisaged payments between $350,000 and $4.2 million depending mainly upon the discounted lifetime earnings of those who died, making a mockery of the idea that all human life is sacrosanct and that we are all in this together irrespective of class (by this actuarial standard, the lives of Afghan civilians, many more of whom have now died in the bombing than died in the World Trade Center, would conveniently be worth almost nothing, since most of them were living on less than a dollar a day). The *New York Times* reported early in December that more than a million people in the metropolitan region were now relying on soup kitchens and charity meals in order to have enough money to pay their rents and avert the fall into homelessness. Nevertheless, homelessness became more and more evident throughout the city. None of this was consistent with the newfound narrative of "American values" or of empathy with the plight of New York. It betokened, in short, a return to that dog-eat-dog American way of life that had prevailed before September 11.

But it was now considerably worsened by the onset of recession.

Consumer confidence was then seized upon as the key to revival. We were suddenly urged to reinstate all the old values, to shop until we dropped, to travel the world as if nothing had changed (even though the national guard was patrolling everywhere, traffic was being stopped and arbitrarily searched, and the lines at airport security were often chaotic), and to "normalize" our lives in abnormal times. Just to drive home the point that this really was a return to business as usual, the president finessed his promise of aid to New York and cynically began to use his popularity to try and ram through even more tax cuts for the corporations and the rich, to drill in the Alaska wildlife refuge, to curb civil liberties and spy on domestic opposition groups, to get conservative judges confirmed, and generally to promote his right-wing agenda at home and his unilateralism abroad ("You are either with us or against us" was the slogan). Freedom was to be curbed to preserve freedom. Bush even turned salesman, letting clips of his speeches be used in TV ads for the travel industry. Giuliani likewise reverted to type and sought to subvert what passes for democracy in New York City by trying to extend his stay as mayor beyond the legal term limits (though his image was by then so burnished it was hard even for him to tarnish it). Tax receipts were in decline, and the city budget was in serious trouble. Cuts in public services were announced. As the economy soured, crime came back with a vengeance, but the police were now so busy protecting potential terrorist targets that street patrols almost disappeared. By mid-December the Manhattan rental market (both residential and commercial) was in serious decline as well.

Locally the awkwardness as to what to do is palpable. New York desperately seeks the return of tourists (after financial services, its most important industry). It even erected a viewing stage at Ground Zero so people could view the scene of the crime. But the many who initially came left so sobered they could not stomach the frivolity of shopping or even going to a Broadway show. By the standards of spectacle, September 11 had to be close to the greatest show on earth, but New York could not easily capitalize on it as a golden opportunity to gain a monopoly rent on this spectacularly unique sightseer destination (though there are signs that this may happen).[2] Local discourses focusing on innocence and grieving get in the way. Developers, hungrily eying this open site of prime real estate, tread gingerly around local sensitivities in formulating their proposals. But they, too, will probably have their way in the end.

The return to normality by other means depended entirely upon how the economy in general and the financial sector in particular might rebound out of their difficulties. This brings us back to the origins of the recession and the role of September 11 in creating or exacerbating it. It was officially determined in late November that the economy had actually been in recession since the previous March. The speculative dot-com bubble had burst, and overcapacity was evident in many sectors. Manufacturing has been winding down, and layoffs were becoming common nationwide. Airlines and hospitality industries were experiencing difficulties. Financial services were also feeling the heat but had been hanging on by a thread. This was the thread that September 11 effectively snapped. But it is important to understand how the thread snapped, because this affects our view of the options and prospects for economic recovery.

Capital, Marx never tired of emphasizing, is a process of circulation rather than a stock of assets. Looked at from the latter point of view, the damage done to the U.S. economy or even to that of Manhattan on September 11 was relatively minor, no worse than that of the Northridge earthquake in Los Angeles in the 1990s. The loss of life was minor in relation to the New York labor market. This was not what snapped the thread. Viewed in terms of circulation, however, the effect of the strike was much more devastating. Shutting down global capital markets and air travel throughout the United States for several days and even offering

---

[2] On monoply rent, see David Harvey, *Spaces of Capital: Towards a Critical Geography* (Routledge, 2001).

commercial-free television had a huge disruptive effect upon flows of capital. The global effect was substantial, but the hit upon the New York economy was severe, particularly given the closure of bridges and tunnels, disruptions to the communications networks, and the immediate effects upon tourism. All of this would have been enough to snap the thread. Cut the circulation process for even a day or two, and severe damage is done.

But a far more serious problem lay beneath the surface, one that was and is destined to play an even more determinant role. John Maynard Keynes, in his famous analysis of the conditions that accompanied the Great Depression of the 1930s, pointed to the significance of the psychology of the market. What he called (drawing covertly on Marx) the "animal spirits" of the entrepreneur were key to maintaining the dynamics of capitalism. This is so because capitalism is inevitably a speculative affair, and one in which expectations (in these times, of both consumers and producers) play such a major role that they often become self-fulfilling. It was largely the animal spirits within U.S.-based capitalism (or what Alan Greenspan referred to as the "irrational exuberance" of financial markets) that sustained global capitalism and overrode all manner of problems in the 1990s (the bailouts of Mexico and Russia, the containment of the Asian crisis and the chronically depressed conditions in Japan, the insipid performance of European economies). And much of this was centered in New York, with all manner of local effects (huge year-end bonuses on Wall Street, for example). What bin Laden's strike did so brilliantly was to undermine confidence by hitting hard at the symbolic center of the system and exposing its vulnerability. It is precisely on this point that the international perspective proves much more helpful than all the talk of attacks on American values and freedom.

New York City had by the year 2000 so attached its fate to the successes and failures of neoliberal globalization and its financial excesses that it seemingly now has little room for maneuver. But the wind in the sails of the triumphalist neoliberalism of the 1990s had already been flagging. The "irrational exuberance" of financial markets was already winding down. The credibility of neoliberal globalization was already being seriously questioned both outside the system (Seattle, Melbourne, Quebec City, Genoa) as well as within. The New York economy was

heading into difficult times well before September 11, but the attack un-
dermined the two last bulwarks of defense against the free fall into recession—the "animal spirits" of both entrepreneurs and consumers.
Simultaneously it exposed the umbilical cord that now tied the fate of a
whole city's economy to the vigor and rhythms of processes of capital circulation through financial markets. This is what the international discourse reveals so starkly, and it is this which the local and the national
discourses on American values effectively hide.

So where, then, can the New York City economy go? In the immediate future there is no avoiding the fact that we are headed into rocky
times. Even if the global recession turns out to be shallow, New York's difficulties will be deeper than most by virtue of its singular attachment to
financial services as its dominant economic base. This is the case because
there are limits and dangers to excessive financialization. To begin with,
almost certainly any economic recovery will have to restore the balance
between the real economy of solid production (of which there is precious
little left in New York, except of the sweatshop variety) and the mere trading of fictitious assets through financial instruments. Systematic economic diversification is not an impossible aim for the city, and this may
be necessary for survival. And it is here that the ingenuity and talents of
the other New York, the immigrants, may well be as crucial to the future
as they have been in the past.

But there is another even more troubling aspect to this. Fernand
Braudel, in examining the long *durée* of capitalism's historical geography
since its origins in the city-states of Renaissance Italy, underscored the
importance of financial expansions as a "sign of autumn" in distinctive
phases of capitalist development led by some hegemonic power. When finance capital took over and dominated all the activities of the business
world, this was, Braudel observed, a sure sign of maturity of an existing
system and its incipient replacement by another.[3] Giovanni Arrighi
amasses considerable evidence for this proposition, going on to note how
the golden glow of periods of financialization and the riches amassed

---

[3] Fernand Braudel, *Capitalism and Material Life, 1400–1800* (Weidenfeld and
Nicholson, 1973).

thereby lull dominant superpowers into a complacency that ill fits them to face the turbulent competition and confrontational geopolitics that lies ahead.[4] The twentieth century was unquestionably "the American century," and New York was the unchallenged center of that hegemony. But New York has also been at the center of the last thirty years in which finance capital has come to dominate all the activities of the business world. In years to come we may learn to see September 11 as the first shot in bringing that particular world to an end. If so, then New York—like Venice, Genoa, Amsterdam, and London in previous centuries—will undoubtedly survive, but will have to adjust to a very different status of subservience rather than mastery of the global economy of capitalism (always presuming, of course, that no alternative mode of production rises up to take capitalism's place). And in that discomforting scenario we may well look back on the brief rediscovery of alternative values in the wake of September 11 as a source of immense strength to brave the difficulties to come.

---

[4] Giovanni Arrighi, *The Long Twentieth Century* (Verso, 1996).

Insecurity by Design

A VERY TALL BUILDING absorbs a plane and collapses after 105 defiant minutes, having watched its twin suffer the same fate. Everyone sees it. Again and again. It captures every eye and ear in stunned amazement. When the towers fell, the world shook. Nobody could accept what they saw. Such a vertical drop seemed impossible. And no amount of analysis of the mechanics of the collapse, the simple way the attack was carried out, or the strategic mission of the attackers can ease the incredulity. The event remains unbelievable, surprising even those who initiated it.

People turn to architects for answers. Surely those responsible for shaping structures, those whose discipline has been discussing the comforts, pleasures, and mysteries of buildings for thousands of years, could help explain the meaning of this traumatic event. One of the first things to rise in the aftermath was the normally marginal figure of the architect. Emerging from relative obscurity, architects were featured in magazines and extensively quoted in newspapers. They appeared on television talk shows and were interviewed on news broadcasts. Every branch of the field was mobilized. It was hard for any designer or critic to remain thoughtfully silent. Indeed, their public statements continue to multiply exponentially. There are endless drawings, models, maps, conferences, think tanks, exhibitions, Web sites, publications, and educational programs. Yet so little is said in the end. Everyone pretends to find in the event the clearest evidence for what they have been saying all along. Scratched records get yet another spin. Mainly it's a kind of disciplinary therapy, a reassertion of the traditional figure of the architect as the generator of culturally reassuring objects, an ongoing denial of the fact that architects are just as confused as the traumatized people they serve.

Why did our seemingly hyperaware and congenitally paranoid world become so shocked? Not because of the number of people killed. Such numbers are tragically all too common on a planet routinely tormented by starvation, war, disease, and genocide. Nor is it simply the terrorist assault on a large metropolitan building. Buildings are constantly being targeted. In a sense, New York simply joined a long list of global capitals whose structures are attacked: London, Buenos Aires, Madrid, Paris, Rome, Berlin, Moscow, and so on. And the United States was already familiar with the deadly collapses of its federal office building in Oklahoma, embassies in Nairobi and Dar es Salaam, and barracks in Beirut. Yet while the front pages of newspapers regularly feature the lethal rubble of flattened modern buildings, none of these collapses stimulated any kind of debate about architecture. The only design response was the development of security perimeters around key buildings to block the access of bomb-laden vehicles, a change that never attracted the interest of architects, despite their notoriously fetishistic obsession with the even the most minor of details. Only with the destruction of the World Trade Center do the designers and critics swarm down like vultures to pick over the site of collapse, unable to offer much because the ancient intimacy between architecture and violence has for so long been off-limits in their discussion. Indeed, "ordinary" people have time and again shown themselves to have a greater depth of understanding of what it meant to lose the Twin Towers and so many of their occupants.

To grasp the event, we need to appreciate the intense fantasies people have about buildings in general and the twins in particular. The issue is one of fundamentalism rather than nuance—not just the fundamentalism of those who launched the attack and those who responded, but the fundamentalism of our most basic beliefs about buildings. To begin to understand the depth and complexity of reaction, we need to go back to the simplest level.

In the simplest terms, buildings are seen as a form of protection, an insulation from danger. They have to be solid because their occupants are fragile. Keeping the elements and enemies out, they allow bodies to have a life. To be hurt by a building is unacceptable. Even the most minor fragmentation of a structure is front-page news. If a piece of a façade falls to the street and slightly injures someone, it will receive more press than most murders. And fatal collapses are international news—death by architecture is intolerable. Furthermore, buildings are traditionally meant to last much longer than people. It is the sense that buildings outlive us that allows us to have a life. Shelter is as much emotional as physical. Buildings shelter life by sustaining a collective sense of time, a form of cultural synchronization. Architects craft time when they craft space. Buildings act as a reassuringly stable witness of whatever we do by surviving longer than us and evolving more slowly. To lose a building is to lose not simply an object that you have been living in or looking at but an object that has been watching over you. And when our witnesses disappear, something of the reality of our life goes with them. People are really grieving for themselves when they grieve for buildings.

This sense that our buildings are our witnesses depends on a kind of kinship between body and building. Not only should buildings protect and last longer than bodies, they must be themselves a kind of body: a surrogate body, a superbody with a face, a façade, that watches us. Western architectural theory began with the argument that the beauty of a building resonates with the supposedly God-given beauty of the body. Architecture evolves as our thinking about the body evolves. Or more the other way around, we use buildings to construct an image of what we would like the body to be. However strange we might find our own fleshy container, thoughts, and emotions, our architecture presents a unified image. Buildings are thereby credited with considerable representational force. This force can be seen in the everyday notion that the place where you live continues to represent you when you are hidden within it, away, asleep, or dead. Like your clothes, your building projects some kind of stable image regardless of what you are doing or feeling.

Terrorists know this, have always known this. They play with these basic fantasies about architecture, wounding buildings as often as people.

Damaged buildings represent damaged bodies. And it is the representation that counts. Terrorism is not about killing people, but about dispersing the threat of death by producing frightening images. Particular sites are targeted to produce a general unease. If you can identify with the target, then your own buildings become unsafe, and every body becomes vulnerable. This tactical use of images of assaulted buildings plays with precisely the representational capacity of buildings that architects have devoted themselves to for millennia. In this, the terrorist shares the expertise of the architect. The terrorist is the exact counter-figure to the architect. Indeed, architects can learn more from the feared criminal than the celebrity designer, more from threats than affirmations. The terrorist mobilizes the whole psychopathology of fears buried beneath the architect's obsession with efficiency, comfort, and pleasure.

The attack on the towers was an extreme yet textbook example. What was unique was the size of the audience and therefore the size of the threat. Symptomatically, the video statement by the structural engineer Osama bin Laden refers to striking the "softest spots" of America, its "greatest buildings," and not the people within them. The real threat is to the architecture, or rather to an architecture that represents a much wider population than its physical occupants. In a classic sense, the targeted buildings represent the bodies of a global constituency, assuming the humanity of all those who watched. Again and again, the towers are described with the same terms used for suffering people: from George Bush saying "the evildoers . . . have hurt our buildings," to the repeated use of expressions like "wounded buildings," "victimized buildings," "tortured structures," "death of the towers," and "death of the twins." In the grieving for those who died, there is also grieving for the buildings themselves. In all the improvised memorials and media coverage, images of the towers' faces share the same space as images of the victims' faces. The buildings became victims, and in so doing victimize those who watch them suffer.

If everyday cultural life makes an unconscious association between body and building, it is enormously frightening when the confusion becomes literal. The devastating spectacle of September 11 was a simultaneous destruction of body and building and the distinction between them. "He became part of the building when it went down," as one distraught parent lamented. The buildings became lethal elevators, dropping in on themselves at the same speed as any object free falling in the air. No resistance. 415 meters of structure compressed into 20 meters of rubble in 10 seconds. Generating a level of energy comparable to nuclear blasts or volcanic activity. Buildings and bodies were instantly compacted into an extraordinarily dense pile or dispersed to the wind. Many of the occupants were instantly "rendered into dust," as the medical examiner's office put it. The resulting cloud blanketed the bottom of the island and was blown to all corners of the city. We literally breathed our architecture and neighbors. The bodies themselves were mainly lost, and even the number of victims stayed a mystery. The few bodies that were found were kept invisible. Despite all the intense and endless media coverage, no bodies were shown. No broken, bloody, burned, or fragmented people. Just the desperate flight of those who could choose to jump and the shocked, dust-laden bodies of the survivors.

The watching world became endlessly captivated by the twisted and pulverized remains of the structures, at once the symbolic representative and literal condition of the people they protected and then destroyed. It was an unprecedented blurring of traditional distinctions. In lieu of remains, family members received small boxes of the dust. Millions streamed downtown to look, partly driven by the voyeuristic compulsion to see the site of such a huge traffic crash and partly like loving family members of the buildings themselves who needed to see the actual body of the buildings to accept their loss.

The towers had of course been designed to produce such a global audience. The whole point was for them to rise up above the city at the end of the island facing Europe to capture world attention. Which they did. They were the centerpieces of billions of images. More postcards were sent of the towers than of any other building in the world. These may well have been the most famous buildings of all. And what was constantly looked up at by so many was also looking back. Whether we were on the streets of Manhattan or watching TV in living rooms on the other side of the world, the Twin Towers had an eye on us. And more than two million people a year came from everywhere to stand on top of them and see what the towers saw. Pressing their eyes to the glass between the narrowly spaced columns, literally putting their heads inside the depth of the façade to share its view. But there now is a palpable sense on the island of having lost a crucial witness that could see you wherever you were: an architecture of image that was understood, and enjoyed.

This popularity was never understood by architects, who are being asked to talk about buildings they never embraced. Indeed, the Twin Towers were mercilessly slammed by architectural critics, particularly those who led the support for so-called postmodern architecture. For them, the towers personified the inhumanity of modern architecture. Ironically, the critics spoke in the name of popular sensibility. The defense of populism in architecture repeatedly assaulted what would turn out to be some of the most popular buildings of all. But in the end ordinary people simply had stronger feelings about the buildings than architects, to whom they rarely listen anyway. The Twin Towers played a much bigger role in everyday experience than in architects' discussions. And the deeply felt affection for these buildings cannot be casually dismissed as the delusions of an exploited public under the manipulative sway of corporate image-makers. At some level, an extraordinary identification with the buildings took place that exceeded the expectations of both the boosters of the project (whose only real interest was real estate speculation) and the architectural critics (whose only real interest was to promote an alternative aesthetic model that would be used in subsequent real estate speculations). Precisely because their brutal scale didn't fit into their surroundings, they perfectly belonged in a city of refugees and misfits of every kind, the city that is at once the most and the least American. People experienced the buildings not as part of some distant power but as an intimate and tangible part of the domestic life of a dispersed global community.

The key symbolic role of the World Trade Center, the rationale for both its design and is destruction, was to represent the global marketplace. In a strange way, supersolid, supervisible, superlocated buildings stood as a figure for the dematerialized, invisible, placeless market. In this, architecture was surely a vestigial symbolic system, as demonstrated by the fact that the markets reopened within days after the attack. Even Cantor Fitzgerald, the securities firm that lost more than 70 percent of its 1,000 workers when the towers collapsed, had the world's biggest electronic bond market back on line within two days. Supposedly fragile digital patterns have long assumed the solidity once associated with buildings. Electronics has taken over from architecture as our primary witness and storehouse of collective memory, allowing buildings to assume new roles. Not by chance were those who first described the architectural role of electronics in the 1950s and 1960s also those who insisted that buildings should become as expendable as dishwashers and toasters. Yet the traumatic reaction to the loss of the towers themselves, beyond the terrible loss of life, shows that even the vestigial system of architecture had more force than anyone expected. Still, it is as yet unclear what it means to threaten a building in an electronic age.

We need to look more closely at exactly how this seemingly outdated representational system worked. The World Trade Center was a hyperdevelopment of the generic postwar corporate office tower. The corporate building provides a fixed, visible face for an unfixed, invisible, and carnivorous organization. Typically, there is no sign of the company on the building. The lack of a literal sign becomes the sign that the corporation is nothing more than an open network. The building's anonymous gridded façade is like an empty classification system, a filing cabinet in which anything could be placed. The façade says: "I am ready to organize anything," "I am here but I could be anywhere else," "I have this shape but I could have any other," "What counts is the relentless open weave of my surface." As is written into the very word, the "corporation" is an abstract body, a corpus composed of many bodies networked together into a single organism. It is an invisible collective network that may be made visible on a particular site by a building. If our default setting is to see buildings in terms of our bodies, the corporate building is a kind of bodiless body. The corporation veils the actual bodies of those whom it networks together and controls from afar and even those who carry out that control. The corporate building is nothing more than a screen that conceals the body. The occupants of such a building are irrelevant.

Indeed, the corporate building is never simply occupied. The shiny glass "curtain wall" is not a showcase revealing the workers inside. During the day, reflections usually render the interior mysterious. At night, when the interior becomes visible, it is the horizontal grid of fluorescent lights visible through the vertical grid of the façade that is put on display, not the workers. The fluorescents, which are never turned off, are more important than the workers. Not by chance is the corporate office building one of the first building types to be routinely photographed at night by architectural publications. The corporate building is never more than a certain kind of façade. Of course, there are internal spaces occupied by workers, and a whole range of tactical design strategies for accommodating them, but this remains subordinate to the polemic of the outer screen. The veiled workspaces simply extend the logic of the façade, with suspended ceilings veiling all the ducts and wires, screening away the guts of the building. At the ground level, the façade is thinned down to the absolute minimum, and the resulting walls of glass invite the eye and body in. But all that is revealed is a huge lobby space that usually continues the pattern and materials of the exterior plaza. What promises to be the interior turns out to be an exterior bathed in eternal artificial sunlight and inhabited only by the elevator core. The elevators are the only substantial objects. If you look up, you don't see the solid underbelly of the building's main volume but an artificial sky, a continuous, brightly glowing, horizontal surface that veils even the shape of the fluorescent tubes that activate it. The hidden office floors above often have the same kind of singular glowing ceiling. And even when the fluorescents are discrete fixtures arrayed across an opaque surface, workers don't inhabit rooms but are distributed across a "landscape." In the most fundamentalist sense, the corporate building has no interior.

The Twin Towers took this to the limit, perfecting the logic of the "neutral" screen, stretching it to the clouds and exemplifying the culture of the invisible body. Unique 18 $3/4$-inch-wide aluminum-clad steel columns defined the face, held apart by just 21 $1/4$ inches of glass. The parallel stripes of metal and glass blended into each other, producing a seamless veil of ambiguous materiality. The towers were at once sealed yet porous, intimidatingly heavy yet floating. The screen became its lightest at the base. In the defining detail of the design, each successive group of three columns was smoothly grafted into a single slender one that lightly touched the ground, dissimulating the structural role of the whole screen and drawing the plaza seamlessly into the lobby (figure 1). The twins

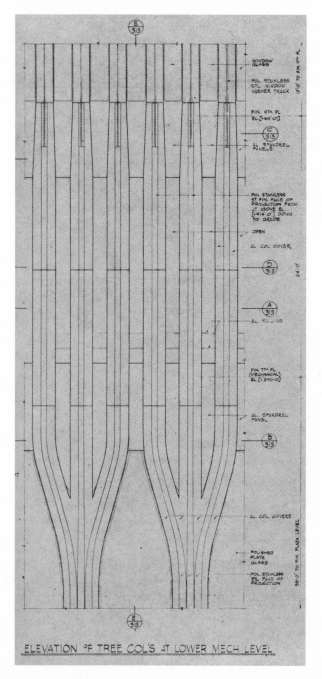

Figure 1. Elevation of "tree" detail between the 6th and 9th floors. *From the architect's set of working drawings for the North Tower.*

were at their most beautiful at night as a complex pattern of lights hover-
ing above the city, framed by the sky-lobbies and observation levels which
turned into thick dark bands. The lights simulated occupation, but not
by people. Indeed, the more empty the building, the more animated the
façade. For the first decade, 23,000 fluorescents always remained at their
posts, working around the clock. There were not even light switches in
the towers until 1982, when the endless stacked layers of light started to
fragment into an elusive and seductive figure.

The two mysterious, gargantuan shafts connected a vast, crowded,
horizontal slab of shopping and eating in the ground to equally crowded
platforms for viewing, eating, and drinking in the sky. A hundred thousand visitors passed through each day. Below the shopping level was a
physical communication hub that radiated multiple underground rail systems. Above the viewing level was an electronic communication hub that
radiated television, telephone, microwave, and radio. The result was a
density of bodies in the spaces of consumption below and above, framed
by communication networks binding the structure to the city and the rest
of the planet. In the space of production in between the two shafts, the
body simply disappeared. Tourists rocketed through it in America's largest
and fastest elevator, suddenly aware of the inside of their own flesh but
oblivious to the spaces and people that surrounded them.

Even the workers did not simply enter or appreciate the space. Isolated in their express elevators, they went to separate sky-lobbies from
which they took local elevators to their respective floors. Nobody knew
who was working there. Even the workers only knew the people on their
own floor or those who happened to be in the same one of the 208 elevators at the same time or passing through the same lobby. One of the consequences of the collapse was finally to reveal who was occupying the
building—the vast range of nationalities, fields, careers, and incomes.
Compounding the already extraordinary levels of identification involved,
the demographics of the workers within the towers turned out to be remarkably similar to that of those who watched their collapse. Yet we
never hear about the spaces they worked in. All that counts is what they
saw out the window, or the atmosphere in the elevators and stairwells as
they escaped. Of course. It is a matter of survival, not aesthetics. Yet so

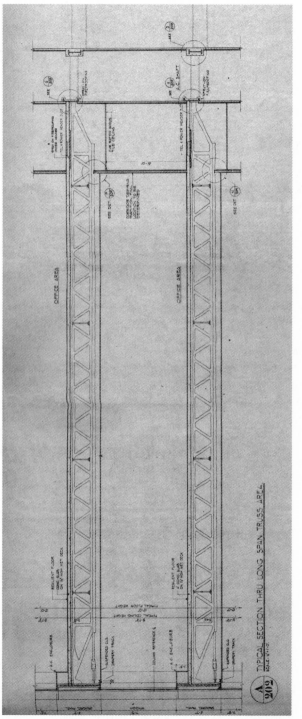

Figure 2. Section showing the uninterrupted slice of office space between the elevator core (right) and the façade (left). *From the architect's set of working drawings for the North Tower.*

much of the subsequent discussion has been carried out in strictly aesthetic terms. And in the deluge of countless images of the building in loving reaction to their unbelievable disappearance, it was symptomatic that not one showed the interior of the workspaces.

The design of these hidden spaces was celebrated in the press as highly innovative when the buildings first went up. The invention of sky-lobbies had freed up an unprecedented amount of useable floor area. Concentrating all the structure into a tight ring of huge columns around the elevator core and another heavy ring around the outside of the building made the workspace column-free. The space became simply an 8-foot, 7 $1/2$-inch-high horizontal slice of controlled air sandwiched between the uninterrupted surfaces of the merely 4-inch-thick concrete floor and the even thinner suspended ceiling (figure 2). The brutally efficient slice ran from the inner service core to the outer glass, with just a 15-inch-high air conditioning duct acting as a slightly defensive window sill. Starting with their own choice of lighting pattern, each of the huge number of tenants could organize their slice freely and differently, setting up a whole array of different relationships to the famous façade, a heterogeneity that was masked by the unified exterior and only subtly implied by the different intensities and rhythms of the light pattern at night. In all the heartfelt expressions of empathy for the trapped workers and all the sudden, sustained attention to architecture, remarkably little attention was given to what kind of spaces these actually were. It is as if everyone tacitly understands that the towers did not have an interior as normally understood.

The towers had no front, back, or sides. Each face was the same. Furthermore, the buildings had no depth. There was no simple view of them in perspective. The smaller buildings ringing their base blocked the view upward, and the windswept plaza on top of the shopping mall was typically deserted. The buildings were meant to be seen from a distance, as pure façade. They were made to seem flat, as can be seen in the architect's original renderings. Each typical view, the cliché embedded in a worldwide consciousness, had two seamless screens side by side: each a shimmering blend of aluminum and glass, subtly modulated by the Gothic- and Arabic-inspired details of the stretched floor heights at the bottom and the top, and the even subtler change of dimension produced by two sky-lobbies. The expressionless face of the mysterious workspaces acted as the default, while a minimal play was reserved for the spaces with the maximum movement of bodies. The curving grafting of the columns at the base could be read from a distance, but the details in the higher floors were so discrete that only their effect was visible, and then only just. In the two sky-lobbies and their adjacent double-height mechanical floors (where glass was replaced by recessed walls of louvers for air movement), the columns smoothly became wider for a while and projected out from the rest of the face by an excruciatingly subtle 7 inches (figure 3). In the

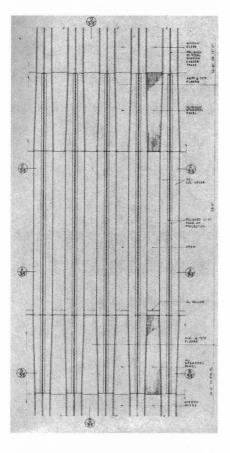

Figure 3. Elevation showing the subtle widening of the columns between the 75th and 77th floors. *From the architect's set of working drawings for the North Tower.*

observation floors and their adjacent mechanical floors, each second column was seemingly split into three strips, with the two outer strips curving over to join the adjacent column before returning four floors later (figure 4). The strips gradually protruded out from the face by a remarkable 5 inches before their return. A sheer face climbed 110 stories above the city, only to have the aluminum trim precisely detailed to bulge out ever so slightly beyond it just before the top. These quiet moves produced the effect of four subtle bands across the face of the monolith that would appear slightly darker or lighter depending on the light. What was unique is that the façade was so strong—the seemingly decorative face actually had the greatest strength. Setting the glass back in the recesses between the huge columns meant that the colors and texture of the face changed continuously with each new angle of view or sun—a minimalist composition that achieved a maximum array of effects. And a remarkable statement for which the architect was never fully acknowledged.

The Twin Towers were a pure, uninhabited image floating above the city, an image forever above the horizon, in some kind of sublime excess, defying our capacity to understand it. The unfathomable trauma of their destruction simply deepened the mystery. And despite the earthshaking intensity of the collapse, the dust finally settled to reveal large sections of the façade improbably left standing, the whole spirit of the building encapsulated in a lonely porous screen whose subtly grafted curves may well have become the most famous architectural detail in history.

The eventual demolition of this poignantly defiant screen was itself foolish and painful. There was an obscene haste to remove all the traces and rebuild in a desperate attempt to fill the void in so many hearts and bank accounts. But the question of how to replace nine million square feet office of space is irrelevant. If anything, the issue is how to replace the more than two million square feet of façade—those vast, uncannily duplicated screens. When the facades came down, the faces of the invisible occupants who were lost came up, filling the vertical surfaces of the city in pasted photocopies and covering the surfaces of televisions, computers, and newspapers all around the world. They

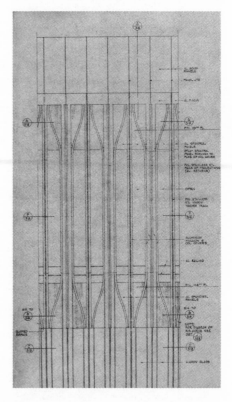

Figure 4. Elevation of crown detail between the 108th and 110th floors. *From the architect's set of working drawings for the North Tower.*

formed a new kind of façade, a dispersed image of diversity in place of the singular monolithic screen—each face, each personality, each story, suddenly in focus. In contrast, survivors were covered with dust, all differences between them concealed by a uniform coating, screened by a thin layer of the building. The old façade still at work. It was precisely those who were missing, those the buildings did not protect, who had their horrifying disappearance marked by a sudden visibility. When architecture rises again, it will likely rebury what was exposed. Another defensive screen will be placed between us and our fears.

This new screen, even, if not especially, that part of it devoted to "memorial," will insulate us from what happened. A city that was able to so completely forget that a third of it was destroyed by a deadly downtown fire in 1776 ("a scene of horror beyond description," said the newspaper of the day) and forget that a quarter of it (seven hundred buildings) was destroyed by another downtown fire in 1835, will be able to forget this latest trauma remarkably quickly. The whole financial district that acted as the site of the latest catastrophe was itself entirely built within a year of the 1835 fire—the first architect having been hired only a day after the fires were put out. Surely a city shaped only by greed will once again find ways to profit quickly from its pain. Appeals to memory and solidarity will be but excuses for multiple forms of local and global restructuring. New shapes of building and social control are easily promoted in the guise of healing the physical and psychological wounds.

Before any such talk about rebuilding, we should have tried to learn what it is to grieve for a building. Indeed, there should have been a patient attempt simply to understand exactly what happened. After all, what occurred is not simply the tragic loss that we can point to, no matter how dramatic and clear-cut it seems. Nothing was easier to point to than the Twin Towers and their collapse. Yet amidst the obvious horror, there is another level of trauma that is even more challenging because we are unwilling to acknowledge it, let alone comprehend it. For what might be really horrifying in the end is precisely what was already there. The collective sense that everything changed that morning may have more to do with no longer being able to repress certain aspects of contemporary life. Things that we have been living with for some time were disturbingly revealed. The everyday idea that architecture keeps the danger out was exposed as a fantasy. Violence is never a distant thing. Security is never more than a fragile illusion. Buildings are much stranger than we are willing to admit. They are tied to the economy of violence rather than simply a protection from it.

When the design of the twins was first revealed in 1964, the architect said that they would be a physical manifestation of "the relationship between world trade and world peace." Did we really think that the emergent forces of globalization were so innocent, and that architecture could embody that innocence? Did we really think that buildings could be in any way neutral? Or did we just agree to pretend? The rationalizations of the rebuilding are just as naïve—and just as successful. Everyday life will isolate itself from lurking nightmares. Once again, business will appear to be separated from memory; a clear prophylactic line will be drawn between "memorial" and routine industrialized spaces for offices, shops, or residences. We will act as if memory itself has not been thoroughly industrialized, that a certain kind of light gift-wrapped remembering is big business today because it dissimulates the ongoing heavy forgetting. The famous panoramic view from the clouds will quickly be replaced by an even more famous close-up. The radical doubts will be reburied. We will again pretend to understand the structures we occupy and observe. The only challenge will be to select the collective forms of denial. And in burying our fears so earnestly, we also bury our pleasures. Architecture will be neutralized and returned to the background. Some architects will have collaborated in this removal of passion from their own domain, but even they will soon fade from our monitors.

Architects are in fact filled with doubt, and often share it when they passionately discuss their designs amongst each other, but they are called on to exude confidence in public. If our buildings are meant to give us confidence, their producers apparently have to embody it. But if architects are not used to bringing their doubts about the status of buildings into public discourse, they are unable to contribute to the much-needed discussion of architecture's intimate and complex relationship to trauma. All they can do is once again collaborate on the production of images of security, comfort, and memory. Once again architecture will rise in the face of crisis, for in a sense architecture is always driven by the need to bury trauma. The embarrassing truth is that the traditional architect is empowered rather than challenged by such events. Architects are in the threat management business. For all their occasional talk about experimentation, they are devoted to the mythology of psychological closure. But the only architecture that might resist the threat of the terrorist is one that already captures the fragility and strangeness of our bodies and identities, an architecture of vulnerability, sensitivity, and perversity. Ignoring this, architects will unwittingly get on with the job of making the next targets.

The Janus Face of Architectural Terrorism: Minoru Yamasaki, Mohammed Atta, and Our World Trade Center[1]

NEARLY TEN YEARS AGO, in what seems from today's vantage point like a relative age of innocence, I began assembling the materials for a book on New York's World Trade Center. My strategy concentrated on observing these buildings as artifacts of a social moment that, in the developed world, coincided with the transition from industrial to information age values. For New York City, the rise of the WTC in the late 1960s marked the threshold between a mixed economy and an emergent monoculture of finance, insurance, and real estate. In their immensity and formal relation to one another, the towers autographed the skyline with an emblematic portal between these eras. They also announced the culmination of elite regional planning strategies several generations in the making, and celebrated, even as Fortune 500 companies fled the city, the emergence of corporate New York triumphantly astride the ruins of manufacturing and lesser commerce.

First proposed by the banker David Rockefeller in the late 1950s as part of a broader Lower Manhattan real estate scheme, the towers were planned and built during the 1960s and seventies by a public agency, the Port Authority of New York and New Jersey (PA). As they rose at the Hudson River's edge, the authority's director, Austin Tobin, described them as

---

[1] An earlier version of this essay appeared on <www.opendemocracy.com>.

a "vertical port." Though Tobin's assertion was disingenuous at best, it is true that the advent of the WTC coincided with the eclipse of New York as the world's premier seaport. The excavations for the foundation of what the PA called "the first buildings of the 21st century" literally buried the piers at the southern edge of Manhattan and ended three hundred years of maritime culture there.

Though my research centered on tracing the political and economic trajectories that gave the towers their distinct form, I also found myself fascinated by the language and thought patterns of the men who had not merely imagined these buildings, but engineered them and harnessed the will of thousands of others to bring them into material existence. I was powerfully struck by the abstract nature of the planners' and architects' thinking, their willingness to reduce lived actuality to a set of disembodied quantities. And so I wrote:

> You need only to stand for a moment in Austin Tobin plaza to become immediately and keenly aware of how [architect Minoru] Yamasaki's abstract sculptural ethos achieved a kind of chilling perfection in his World Trade Center design. Here you find yourself in the presence of two monumental structures whose formal relationship gives us no indication of their purpose or intent. You know they are office buildings, yet their design makes it nearly impossible to imagine that they are full of *people*. It is at this point that—even without invoking the optical trick of standing at a towers' corner and looking upward—you realize the trade towers disappear as sites of human habitation and reassert their power at the level of an esthetic relationship. And it is through recognizing this process that you may become uncomfortably aware of a kindred spirit linking the apparently polar realms of skyscraper terrorist and skyscraper builder.

> This analogy between those who seek to destroy the structures the latter thought it rational and desirable to build becomes possible by shifting focus momentarily to the shared, underlying predicate of their acts. To attempt creation or destruction on such an immense scale requires both bombers and master-builders to view

living processes in general, and social life in particular, with a high degree of abstraction. Both must undertake a radical distancing of themselves from the flesh and blood experience of mundane existence "on the ground." Gaston Bachelard observed in *The Poetics of Space* that attaining such a state requires one to manufacture a "daydream": a reverie in which one observes others as they "move about irrationally 'like ants.'" Separated from "the restless world" of the here and now, the daydream world offers up the "impression of domination at little cost."

For Bachelard, the design of the tall building demands, as the price of its extreme verticality, the sacrifice of a "dream cellar." The skyscraper fails to make room for the volatile urges that raised it to be explored, acknowledged and integrated. It remains, in Bachelard's term, "oneirically incomplete"—robbed of space for the language of the unconscious. Thus our city of towers stands condemned to communicate only one side of the dialogue—it transmits messages of a "purely *exterior*" value alone.

Through building and inhabiting our towers, we push ourselves toward a break in connection with the stuff of our own humanness. For Bachelard, the skyscrapers' elevators "do away with the heroism of stair climbing. . . . Everything about [them] is mechanical and, on every side, intimate living flees." For the terrorist and the skyscraper builder alike, day-to-day existence shrinks to insignificance—reality distills itself to the instrumental use of physical forces in service of an abstract goal. Engulfed by their daydream, they are "no longer aware of the outside universe."[2]

When these words were published in 1999, several reviewers—including some who praised my book—took me to task for making such a comparison. Perhaps they found the equation facile or else were not prepared to allow that the narrow separation between the two mind-sets made them uncomfortable. So on this point in particular, I found myself feel-

[2] Eric Darton, *Divided We Stand: A Biography of New York's World Trade Center* (Basic Books, 1999), 118–19.

ing that I had climbed out on a very long limb where I could expect precious little company.

Then arrived, literally on my doorstep, a hideous confirmation of my
thesis in the form of an article in the October 10, 2001, *New York Times*.
The story concerned the transformation of Mohammed Atta from "a shy
young man" into the "mastermind" behind the destruction of the World
Trade Center. It was Atta, apparently, who led the team of hijackers and
himself piloted one of the planes that brought the towers down—though
the *Times*, stuttering through his unhappy Odyssey, could not bring itself
to report him a suicide in so many words. Born in Egypt and graduating
from Cairo University in 1980, Atta was awarded a scholarship from
Hamburg Technical University, and, upon receiving his architecture degree, went to work for a German urban planning firm, Plankontor. There
he "impressed his co-workers with his diligence and the careful elegance
of his drafting." But "instead of becoming an architect or urban planner,
Mr. Atta became an Islamic terrorist."

Based on the *Times*'s biography, a number of comparisons may be
drawn between the lives of Mohammed Atta and the WTC's chief architect, Minoru Yamasaki, familiarly known as Yama. As boys, both were
dominated by autocratic fathers whose professional aspirations for their
sons sometimes took the form of devastating psychological cruelty. Both
Yama and Atta found themselves alienated within their respective cultures. The son of Japanese immigrants, Yamasaki grew up in the Pacific
Northwest, where he endured the vicious racism of the World War II era.
Even in later years, after he had become a successful architect, Yamasaki
faced open discrimination when he tried to buy a house in suburban
Michigan. Mr. Atta reportedly felt alienated in Germany, but even more
so when he returned to Egypt and found that in Cairo, urban renewal
amounted to knocking down poor neighborhoods in order to, as a German classmate put it, "make a Disneyworld out of it." And so he fled
back to Germany and reportedly into the arms of the jihad.[3]

Whatever the value of such psychological profiling, the parallels be-

---

[3] "The Mastermind; A Portrait of the Terrorist: From Shy Child to Single-Minded
Killer." Jim Yardley with Neil MacFarquhar and Paul Zeilbauer, *New York Times*, October
10, 2001, B9:1.

tween Yamasaki and Atta would seem a kind of "DNA match" between members of the same highly disciplined profession exercising their skills at the highest level to opposing purposes. Bluntly put, the Port Authority of New York and New Jersey hired Minoru Yamasaki to erect its towers. A generation later, Osama bin Laden (we presume) contracted Atta to un-build them. Though he cannot be called the towers' client of record, David Rockefeller, self-proclaimed advocate of "catalytic bigness," found articulators of his vision in Tobin and, by extension, Yamasaki. We will probably never know the identities of those who stood in the shadows be-hind bin Laden when he signed on Mohammed Atta.

Examining the nearly forensic interlock between Yamasaki and Atta may be useful in comparing the building of the WTC and its destruction as enactments of polarized daydreams of domination. Whether a master plan entails casting away stones or gathering stones together, the project rests upon the creation of an abstract, quantitative logic that supposes itself to operate on a higher plane than that inhabited by the human ma-terial beneath it. You can package 50,000 people in a 10-million-square-foot office block, accounting for weight and windloads and, as Yamasaki did, proclaim it a "symbol of world peace." Sure, no problem. And at the other end, you can calculate the structural properties of the target, the projectile's velocity on impact, the necessary payload of jet fuel. No prob-lem. You just do the math.

Now, it seems to me that recognizing the downside of our capacity for certain modes of thinking should not lead us reflexively to renounce or suppress them. We might rather, and to our benefit, enfold our abstract and quantitative tendencies within a wider awareness—one that permits bold imaginative leaps yet awakens us from our daydreams when their en-actment pushes us past the point where we can recognize the human form or perceive the existence of concrete human needs.

It seems a particularly difficult task, in the aftermath of the horrific deaths of thousands and the deeply disconcerting failure of what seemed like so much enduring, reliable mass, to contemplate the building of the World Trade Center itself as a destructive act—specifically, an attack planned by the city's oligarchs and carried out with the general consent of its populace. Yet this aspect of the towers needs to be integrated into our historical awareness if we hope not just to rebuild the structures but to reimagine what it is to be New Yorkers and citizens of the developed world.

Until September 11, 2001, the parallel monoliths of the WTC stood hypervisible against the skyline yet remained largely unknown, even to their inhabitants. Less clear was the degree to which the towers were artificial implants in the city fabric: moored in Manhattan schist like any other skyscraper, yet not *of* New York. Beyond their architectural design, much of this inorganic, incongruous quality derived from the peculiar constitution of the Port Authority as a sovereign legal entity whose powers trumped those of the municipality on whose land they were constructed. This political asymmetry allowed for the condemnation of the World Trade Center site under eminent domain and resulted in the legally sanctioned displacement of an entire commercial community—hundreds of people engaged in scores of small and medium-sized businesses collectively known as Radio Row.

Berenice Abbott. Radio Row, Cortlandt Street between Washington and Greenwich Streets, April 8, 1936. Museum of the City of New York, Federal Arts Project.

Some of the most viscerally affecting primary documents I encountered in the course of my research were the PA's own files detailing the particulars of the condemned buildings located on what would become the World Trade Center's footprint. The records made punctilious note of the specifications of the structures to be pulled down and the price paid to the owners, generally far below market value. For one four-story loft building on Greenwich Street, indistinguishable from nearby survivors now converted to luxury condominiums, the PA paid the owner $16,000. Stapled to the cover sheet for each property was a snapshot of the building taken just prior to its demolition. Thus encountered, more than thirty years after the fact, this collection of individual records testified, with a kind of breathtaking directness, not just to the particularity of what had been demolished, but also to the methodical abstractedness with which the agency undertook its mission. Now, given the destruction of the trade center and with it, doubtless, much of the PA archives, who knows what has become of these documents?

This, in any case, was the manner whereby New York City lost jurisdiction over 16 acres of arguably some of its most valuable land and suffered the obliteration of a significant portion of its historic Lower Manhattan streetscape. Expropriation of the property by the Port Authority also meant that what the city received annually in lieu of taxes represented only a small fraction of the amount a commercial developer would have had to ante up. Just before demolitions on the WTC site began in 1966, the New York City Planning Commission, never the most robust agent of democratic decision-making, found its role relegated to rubber-stamping Austin Tobin's *fait accompli.*

Nor, in constructing the World Trade Center, was the PA obliged by law to comply with the New York City building code, an issue which, given the enormity of the structural failure and loss of life, will surely resonate for years to come. What kind of radical disconnect was operating in the minds of those who decided to substitute a double layer of Sheetrock for the standard concrete as fireproofing for the towers' core columns? Less durable, but cheaper and faster. There, in one stroke, went the traditional four-hour window in the firefighters' manual for high-rise blazes. And a sprinkler system—another safety commonplace—was ruled out because of the weight of the water tanks. In the end, New York ended up with twin buildings so tall that, under optimal conditions, it took 45 minutes to descend them on foot. Of course one can conceive of such things, but would it not be pure, enacted domination fantasy actually to build them? How far does a structure have to veer from fundamental considerations of human life and safety before we can recognize it as a manifestation of terrorism?

Today, in the wake of the World Trade Center's violent unbuilding, in the presence of the 1.5 million tons of wreckage the city has been left with, including 200,000 tons of structural steel, 6 acres of marble—or on a more palpable scale, 40,000 doorknobs and 5,000 bathroom soap dispensers—it is possible to grasp the incalculably great price we pay for ceding our political wills to sovereign authorities with big plans for us little people. We start paying it the moment we allow ourselves to be abstracted out of our day-to-day lives and identify as our own the voice that tells us it is our destiny as individuals and a culture to inhabit an empyreal realm where we may possess "domination at little cost." This delusory state was one face of the terrorism built into the World Trade Center. On September 11, the other face became visible with jaw-dropping suddenness.

I am not making an argument for going backward to some presumably more humane urban past. Nor am I saying: "Death to the skyscraper!"—though I think its American incarnation has probably come and gone. What I am advocating is moving forward differently, into a city and a world in which our abstract thoughts become a function of our integrative capacities and where we plan ourselves *into* our structures rather than attempting to "humanize" the distorted results of our, or others', omnipotent yearnings.

A few days after the WTC towers fell in upon themselves, I sat with a friend at our local coffee shop in Chelsea. Already the energies of our vastly powerful nation were driving toward war. Nonetheless, we speculated on his most recent daydream: a futuristic grooved railway—a supply and trade route from New York City, up across Canada, over the Bering Straits, through Asia, into the heart of Europe and south into Africa. Our talk grew animated as we elaborated the idea of a massive, international, public works project, linking a host of autonomous yet interdependent localities. What would it feel like, we wondered, to live in a city at once more deeply connected to itself and to other places on the earth? But since he is an economist and craftsman and I am a writer and a teacher, we let our conversation go at that. Parting to head our separate ways, we agreed that in the near term, New York City should relearn how to build good ships. After all, the harbor is still there. And the city's lifeblood was always the sea.

We humans are born creatures of the earth and air, capable of functioning with our heads in the clouds—so long as our feet remain on the ground. Rising toward the stratosphere, though, we feel we have broken free of gravity. When that narcotic sense of weightlessness possesses us, it is not long before our ascent finds its opposite number in the terror of the fall.

Scales of Terror:
The Manufacturing of Nationalism
and the War for U.S. Globalism[1]

> America, you have it better
> Than our continent that's old.
> You have no crumbling castles
> Nor basalt either.
> Within, nothing daunts you
> In times rife with life
> No memory haunts you
> Nor vain, idle strife. . . .
> —Goethe, 1796

THE WORLD TRADE CENTER catastrophe was a profoundly local event. As the dust settled, rescue efforts turned into cleanup operations, and the debate over reconstruction spilled out, Ground Zero was revealed to be only 16 acres. New Yorkers experienced it as a localized event in real time and space: the planes hitting, the towers aflame, their awesome, inconceivable collapse, acrid poisonous smoke billowing up Broadway, paper floating over to Brooklyn, ash on the pavement. Only later did the constant reruns on television make the catastrophe sensible for us at other geographical scales.

---

[1] An earlier and significantly different version of this piece appeared as "Scales of Terror and the Resort to Geography: September 11, October 7," *Society and Space* 18 (2001).

For when the swirling dust began to settle, the obvious became starkly clear: this was quintessentially a global event. The hijackers were of several nationalities, had driver's licences from yet others, and their most recent residences were in Boca Raton, Hamburg, London, Toronto, New Jersey. The victims also spanned the globe, hailing from an estimated 83 nations, with interim reports suggesting that Pakistan, Mexico, Colombia, and Great Britain probably suffered the highest numbers of casualties after the United States. The targets themselves were international icons: the *World* Trade Center, symbol of global financial power, and the Pentagon, home as much as symbol of global military power. The military response took the form of an "*International* Coalition against Terrorism," while at least one hysterical conservative hyped it as the opening salvo of a "World War III" against Arabs (Thomas L. Friedman, *New York Times*, September 13, 2001). Images of the catastrophe were instantaneously global; New Yorkers reassuring relatives by phone in India, Scotland, Colombia, and Egypt in the minutes after the first plane sliced into the North Tower found that television stations around the world were already reporting the event live. With an extraordinary range of emotions, the world watched in disbelief as a power unanticipated pierced a power that thought itself invincible.

A global event and yet utterly local: how did September 11 become a *national* tragedy? The local and global scaling of this terror has not disappeared, but it has been overshadowed by a powerful nationalization of grief, anger, and reciprocal terror. In the minutes, hours, and days after the planes hit, stunned media executives had no automatic script to work with, and the narratives and images that came across were consequently raw, not just insofar as live images were broadcast of bodies tumbling earthward against the façade of the World Trade Center, but because commentators struggling to make sense of an inconceivable horror had to abandon the safety net of anaesthetic hermeticism that defines their habitual scripts. They were naked, face to face with the horrors of truth. Indeed, in some instances September 11 acted as a sort of truth serum. On September 13, one National Public Radio commentator went so far as to say that the revolutionary analyses of early-twentieth-century Italian socialist Antonio Gramsci had renewed relevance.

Amid the truth wrought by confusion, confusion was also being co-
erced from the truth. Shortly after the planes hit, the ShopNBC television channel suspended programming due to the tragic impact "on each
of us as American citizens." This erasure of multinational victimhood—
several hundred of the approximately 2,700 WTC workers were non-
American—was compounded by the epidemic of flag-waving that
ensued, whether out of patriotism or self-defense on the part of South
Asian taxi drivers, Egyptian cafe owners, and Chinese shopkeepers. The
nationalization of 9/11 was virtually complete by the time Newt Gin-
grich, ex–history professor and previously leader of the House of Repre-
sentatives, advocated on a television talk show that "we" (the U.S.
government) should bomb all "these nations," and thereby demonstrate
"the superiority of Western civilization."

There was little that was automatically national in the scale of these
attacks, however. Both targets were certainly on U.S. soil but it was the
World Trade Center and the Pentagon that were targeted, not the Statue
of Liberty, Disney World, or Hollywood, which are arguably much more
resonant symbols of American national identity.

September 11 was an attack on global economic and military power,
which, in recent years, resides disproportionately in the United States, and
the silences amid the discursive hysteria suggest the angst involved in na-
tionalizing the event. Indignation fastened on the World Trade Center,
while the disturbingly easy destruction of one wing of the country's military headquarters fell from focus. Media and military muteness about the
Pentagon attack highlights the contradictions of rendering this a national-
scale event: on the one hand, the Pentagon attack should presumably provide a vital lever for building the case about an "attack on America," since
targeting the WTC alone could have many other meanings; yet on the
other hand, the deliberate, devastating crash of a commercial jet into the
headquarters of global military power symbolizes the stunning vulnerability of that national power. Highlighting the attack on the Pentagon might
have sharpened the patriotic response, but at the expense of exposing the
entire episode as a stunning national embarrassment.

Not only were the victims quickly nationalized, but also the new enemy. It was a short and fast jump from the suspicion surrounding

ex–Saudi citizen Osama bin Laden and the multinational network Al
Qaeda, to the demonization of the quite demonizable Taliban, to the pro-
jection of Afghanistan in the crosshairs of war. I have seen no poll, but as
war against Afghanistan winds down with minimal success—the Taliban
government routed in favor of fragmented fiefdoms, bin Laden and most
of the Al Qaeda leadership apparently escaped—and the "war against terrorism" scouts Iraq, North Korea, and Iran for further targets, a large
number of Americans take it as axiomatic that Afghanistan was responsible for the WTC attack.

Attacks, harassment, and in several cases the murder of American
Arabs, Indians, and Muslims after 9/11 expressed both the ugliness and
the psychic desperation of the hardening nationalism; some Jews in New
York took to wearing yarmulkes as a means of protection, while news stations reported the day after the attacks that several "Arab nationals" had
been detained in New York. Cultural solipsism and global ignorance fuel
each other: there is of course no such thing as an "Arab national," unless
of course there are also Caucasian, Aryan, Eskimo, or Jewish "nationals,"
but the significant point again is the desperation with which forms of social difference are collapsed into national difference. Long before apparently domestic terrorists visited postal anthrax on government and media
targets, Arabs, Muslims, Sikhs, South Asians—those who simply "looked
Middle Eastern"—had become a form of *social* anthrax in America.

The manufacture of a national scale of response cannot simply be explained as the result of dense media bombardment. Powerful images were
matched by a very practical hardening of national geographies and identities. Less than two hours after the first plane hit, "we" sealed ourselves off
from the rest of the world. The Mexican-U.S. border was unilaterally
closed, then the Canadian border, all U.S. airports were closed and all in-
coming international flights were diverted north or south or back to their
origins. Accusations of a new isolationism, hurled at the U.S. government
after its unilateral abstention from the Kyoto environmental accords earlier in the year and its ruthless efforts to wreck the global conference
against racism, were transformed from political metaphor into
extraordinary geographical truth. Not just the borders but the national
pastime of moneymaking was closed down—capitalism as usual was sus-

pended. The New York Stock Exchange (NYSE) quickly closed (causing domino closures around the world) and remained closed for an unprecedented four trading days; transnational currency transactions were suspended; and in the following days, a six-month-old recession was laid at the doorstep of September 11 terrorism.

The suspension of capitalism as usual, entwined with the anxious nationalization of the event, had the effect of revealing in stark outline the very real fusion (and confusion) of an ideological Americanism with the interests of global capitalism. New York City's mayor, Rudy Giuliani, made the connection concrete when, after the initial shock of the catastrophe had sunk in and the disastrous effect on New York's economy became clear, he appealed to the patriotism of New Yorkers and tourists alike: "go to restaurants, go to shows, go out and spend money." Similarly, when the NYSE reopened on September 17, U.S. Treasury Secretary Paul O'Neill presided over the ceremony, which included not only a two-minute silence for victims but a singing of "God Bless America." Thus were the civic virtues of capitalism and the sureties of religious destiny spliced into the fervid nationalist scaling of 9/11.

Self-imposed or otherwise, the level of press censorship, especially since October 7, when the war against Afghanistan began, completes the dialectic of ignorance and national victimization. This is glaringly obvious in any comparison of U.S. and non-U.S. news sources. U.S. media have cheerfully capitulated to White House requests that the media exercise "restraint" in their reporting of the war, avoiding for "security reasons" the use of "unofficial" sources. Accordingly, one network executive reasoned that showing images of Afghanis killed by American bombs would be "perverse" given the huge toll at the World Trade Center. The defensive self-justification of this logic is as tightly sealed as the nation's airports in the days after the attack. In fact, while the estimate of victims at the WTC has dipped below 3,000, estimates of innocent casualties in the Afghan war are now close to 4,000 and rising, although one would search in vain for this figure in the U.S. media (Seumas Milne, "The Innocent Dead in a Coward's War," *The Guardian* December 20, 2001). But the true perversity here is not just that "our" casualties are judged to be worth more than "theirs"—thousands of dead Afghans airbrushed from history—but that fighting terror and war is surely never a moral numbers game. Innocent Afghan children, women, and men no more deserved to die after October 7 than WTC workers did a month earlier.

With the war only three weeks old, the British *Daily Mirror* could editorialize that "the war against terrorism is a fraud" (John Pilger, "Hidden Agenda behind War against Terrorism," *Daily Mirror*, October 29, 2001): it was disorganized, had few targets, and failed to find bin Laden This from a conservative but honest tabloid in the land of America's major ally. Much more oblique criticism in the U.S. media was generally cocooned by official assurances, although *USA Today* did come out early against the new security state and the illegal, open-ended detainment of "foreign suspects." Still, the net effect of press self-censorship in the United States was not just to neuter domestic opposition but to starch a self-fulfilling nationalist veil of popular ignorance around the United States. Willfully as much as instinctively, the press and government combined to immunize the U.S. populace against the facts and images of a war carried out in the name of U.S. citizens and the horrified responses of people around the world. It is difficult not to recall John F. Kennedy, who once observed, however candidly, that "a nation that is afraid to let its people judge the truth and falsehood in an open market, is a nation that is afraid of its people."

Fabricated ignorance of the world is most powerfully expressed in the plaintive American refrain, "Why do they hate us so much?"—a question whose framing guarantees a fallacious answer. Behind the undefined "they" lurks a racist pigeonholing of Arabs and Muslims, of course, but less obvious is the radical homogenization implied by "us." The "us" here assumes a seamless if simplistic correspondence between the people of the United States and the U.S. government that under any other conditions would be thought ludicrous. Most non-Americans understand very well the distinction between the American people and the behavior of the U.S. government, which is exactly why the WTC and the Pentagon— possibly also the White House, but not the Statue of Liberty or Disney World—were the targets. More generally, the rest of the world understands the United States better than Americans understand them. There is no mystery in this. What happens in or at the behest of the United

States intimately affects people in the rest of the world because of the global political, economic, and cultural power the United States wields. U.S. domestic news is therefore of immediate import, reported world-wide, whereas comparatively little "world news" filters through to the United States precisely because the same political, economic, and cultural power immunizes "us" from everyone else's concerns. It is therefore no accident that the most cosmopolitan nation in the world is simultaneously one of the most insular and uninformed about events beyond its borders.

All of this puts in stark relief the contradiction between assumed geographical privilege and the antigeographical ideology of post–nineteenth-century Americanism in which this assumption of national privilege inheres.[2] Somehow, through nearly a century in which the United States rose to global hegemony, none of the brutal wars—20 million killed in World War I, more than 30 million in World War II, and many tens of millions in other wars on four continents—touched the national territory of the United States. Not since the war of 1812 has there been a significant incursion on the U.S. mainland. No other country has been so immune to the terror that made the twentieth century the most violent in history yet so implicated in it. Nowhere else has a populace had the luxury of deluding themselves that geography is salvation, that geography protects power. If this delusion of geographical exceptionalism is now punctured, the anxious nationalization of terror and war has simultaneously worked to reassert hard geographical boundaries around the United States while framing Americanism *as* globalism. Nowhere was this more vivid than when U.S. troops, having captured a mountainous region of Afghanistan and in a move reminiscent of Teddy Roosevelt on San Juan Hill, staged a highly publicized planting of the stars and stripes. The Department of Defense sensed the contradiction—if this is an international war against terrorism, why is the U.S. flag the appropriate symbol of victory?—and quietly ordered its generals to mute such overt nationalist celebration on Afghan soil.

---

[2] See Neil Smith, "Lost Geography of the American Century," in *Mapping the American Century: Isaiah Bowman and the Prelude to Globalization* (University of California Press, forthcoming 2003).

Much effort went into the reworking of a simultaneously local and global tragedy for nationalist purposes. Ideological preparation for war was the desired payoff, and nationalism *is* the discourse of war under modern capitalism, in which the national state has cornered a monopoly on violence. In this case it is a national monopoly over violence asserted at the global scale. If the "war on terrorism" is a fraud, therefore, it is not simply because the U.S. government has arrogated to itself the right to decide who does and who does not count as a terrorist, nor is it because global race profiling makes Muslims and Arabs and Muslim and Arab countries automatic suspects, along with so-called rogue states, while U.S.-sponsored terror from Colombia to Palestine to Afghanistan itself is exempted as a necessary struggle for freedom. Rather it is because the war on terrorism is not really about terrorism.

The slippage between global and nationalist definitions of terrorism, between universal agreement and narrowly American interests, lies at the heart of the question. In reality, the attacks on the Pentagon and World Trade Center have given the U.S. elite the opportunity to pursue a war conceived as the endgame of globalization. It is a war whose real intent is to establish U.S. hegemony in the Middle East, a power broadly eroded in the 1970s with the assertion of OPEC's influence and the 1979 revolution in the client state of Iran. For its possession of massive petroleum resources, the Middle East is a vital geopolitical region, but this is not a war, as some on the Left have claimed, over oil. Such old-fashioned geopolitical calculations are not entirely obsolete today, but they are secondary. Rather, it is a geoeconomic war to reassert control in the only remaining region of the post–Cold War world that mounts a serious threat to the vision of neoliberal globalization emanating from New York, Washington, and London since the 1980s. Various strands of Islam represent an alternative modernity—not just vis-à-vis the United States but often against Arab states themselves—and "antiterrorism" is a convenient galvanizing ideology for this war. This is the real meaning of repeated calls to move on from Afghanistan, to "finish off Saddam Hussein," attack Somalia, smash the Sudan.

The slippage of global and national perspectives is like looking through a telescope from one end and then from the other. Looked at through the wrong end first, the world is seen at an exaggerated distance from an isolated point: the U.S. elite claims the mantle of global protector over a world too far off to be seen. Looked at through the thin end, with events closely in focus, the world appears much more diverse, and the singular nationality of U.S. global claims—"a global war against terrorism"—is too large for the global space it seeks to inhabit.

There have been three crucial moments of U.S. global ambition. The first came in the quarter-century after 1898 when copious supplies of financial capital, the effervescence of the world's most powerful manufacturing economy, fueled unprecedented international investment; Woodrow Wilson promoted a "global Monroe Doctrine"; and the Council on Foreign Relations was established with the extraordinary ambition of fashioning the country's first liberal foreign policy. The Versailles Treaty and the Depression scuttled that optimism. The second moment followed World War II, when Franklin Roosevelt's new world order offered up a second chance, but the circumscribed geographies of the Cold War, the revolts of the 1960s, national liberation movements, and defeat in Vietnam again frustrated U.S. global ambition. The third moment of U.S. globalism gathered steam in the 1980s, accelerated with the implosion of the Soviet state, and expressed itself in the new neoliberal financialization of the global economy in the 1990s. Insistence by the U.S. government that Afghanistan is just the beginning and that the so-called war against terrorism will take years raises the very real possibility that we will come to see September 11—more accurately October 7, when war commenced—as the fulcrum on which the third moment of U.S. global ambition was adjourned.

Not only is the war itself far more precarious than the U.S. media reports, but it has given a whole new generation of young Muslims a vivid reason for rage and a target that can never be adequately protected. The war in Afghanistan has created more converts worldwide than bin Laden could have dreamt of. Domestically, the nationalist globalism—we are the world?—is much more fragile than the epidemic flag-waving would suggest. With as many as three million Arabs in the United States and five million Muslims, media and the government were quickly obliged to make a distinction between "good Muslims" and "bad Muslims," good

Arabs and bad Arabs, a distinction George W. Bush pushed home during a clumsy visit to a Washington mosque, where he insisted that the coming war against Muslims and Arabs was not a war against Muslims and Arabs.

Insofar as Americanism is now bound up with the distinction between so-called good Muslims and Arabs and bad ones, the definition of American national identity itself becomes fragile. So long as the U.S. elite controls the definition of terrorism, that fragility may not be evident, but definitions are never stable. "Terrorism" long had quite positive allusions, as in Menachem Begin's proud claim in the 1940s, while fighting the British and Palestinians to carve out a Jewish state, that "we are terrorists." The idea of state terrorism was generalized only in the 1970s. A redefinition of terrorism today has the potential to destabilize definitions of American national identity.

Hence the official silence about the identity of the anthrax terrorists—presumed, like bin Laden, to have been trained by the U.S. government—which contrasts sharply with the demonization of "foreign" terrorists. From Timothy McVeigh (also trained by the U.S. military) to the Columbine high school shooters, domestic terrorists are rarely objectified as such. And why else would the White House insist, without clarification, that Charles Bishop, the Florida teenager who flew a plane into a Tampa Bay skyscraper in a "gesture of support" for bin Laden, was not a terrorist? The answer lies in the enormity of the ideological stakes: if Americans are also terrorists, why is the "war on terrorism" exclusively focused on the Middle East or against Muslims? If the extreme selectivity of the U.S. response to terrorism is admitted—if Americans can be terrorists—why does the bombing of Afghanistan, not directly implicated in the WTC and Pentagon attacks, not itself count as an act of gross state terrorism? It all depends which end of the telescope one looks through. Thus some of the first U.S. reporters into Afghanistan in early November 2001 were confronted by angry civilians whose villages had been bombed and relatives killed and who clamored to know: "Why do Americans hate us so much?" The transnational symmetry of complaint speaks volumes.

The notion of a networked global empire freed of geographical boundaries and spatial differences—a decentralized and deterritorialized empire which "establishes no territorial centers of power and does not rely on fixed boundaries or barriers"[3]—has recently come into vogue. In its more conservative versions, this utopianism—literally a spaceless world—stands as an apology for a globalization apparently fueled by the free movement of deregulated finance capital. In its more progressive visage, this notion tends toward postnationalism, postimperialism, and "the network society." Analyses of the Al Qaeda network, about which so little is known, have cited its nonnational yet multinational character as symptomatic of the new spatially fluid definition of power. Whether from right or left, however, it seems to me that the events of, and responses to, September 11 dramatically collapse this utopianism with the same finality as the fall of the World Trade Center. New York is very much an imperial center, and while some borders, such as those of Afghanistan, have been unceremoniously dissolved, others have been dramatically refixed, in particular those around the United States and around U.S. national identity. The spatiality of empire was emphatically reasserted after 9/11 and 10/7 and comes at considerable global cost.

Globalization would seem to require some sort of global state, and it is increasingly clear that the United States covets that role for itself. After Afghanistan, the state apparatus scans the world—defining, judging, sentencing—to determine where and how a new intervention can best clear the decks for its own globalism. More than ever the United States looks like the most dangerous of rogue states. We live in a neoliberal world in which the central contradiction, we are told, lies between the private market and the state. This raises the ironic yet alarming specter of a future in which the struggle for a new globalism will be defined by the contest between, on the one side, private market terrorism of the bin Laden variety, and on the other, state terrorism of the sort promulgated by the United States and Israel, Britain and Russia.

Although other possibilities are largely foreclosed by the fabricated ignorance dished up by a regrouped media—no more Gramsci on NPR—alternatives are nonetheless there. A groundswell of international discontent accompanies the ongoing war, and the fragility of this nationalist globalism is increasingly apparent. One can only hope for and work

---

[3] Michael Hardt and Antonio Negri, *Empire* (Harvard University Press, 2000), xii.

for the bilateral defeat of both the U.S. war machine and religious au-
thoritarians—both using God as an alibi for terror. Neither expresses a viable future. Yet it has to be said that the antiglobalization movement,
surging since the late 1990s, has been dealt a body blow by September
11, one partly of its own making. So entranced were so many in that
movement with the spaceless universality of the new global empire that
they were unable to define the U.S. state itself as a spatially defined agent
of globalism and therefore focus on it as a target for their politics. The
cancellation of the huge antiglobalization demonstration in Washington,
D.C., at the end of September 2001, which was likely to attract 100,000
people or more (some groups did turn up anyway), was in this respect a
disastrous mistake. If indeed the so-called war against terrorism is nothing more than a war for U.S. globalism, then the important point today is
to see anti-globalization politics as simultaneously antiwar politics. How,
then, do we find a way of articulating an antiterrorist politics that has
both private-market and state-sponsored terrorists in its sights?

Meditations on a Wounded Skyline and Its Stratigraphies of Pain

SINCE SEPTEMBER 11, 2001, the skyline of Manhattan is absent its two largest peaks; the loss has disseminated wrenching pain across multiple strata in the life of the city. It left an open wound on the ground that constantly references a past event yet speaks to its lingering presence. The current lack of closure at the wounded site—both the skyline and the ground—also underscores the indeterminacy of what the future holds when the time for rebuilding arrives. How does one approach the reality of the event of September 11 and simultaneously retain it as a critical and reflective activity? If only the wound in the skyline becomes the center of focus and the void in its outline quickly filled in, then the narration of human events related to the site's wreckage will be buried and forgotten too soon. Conversely, if only the injuries are accounted for, the deaths enumerated, the economic losses piled up, then collective pain and memory will remain an abstraction, dissolved into insignificant details and imposed coherence. Contrary to the desire that seeks closure by rebuilding the void in the heart of the financial district, and in spite of the need to recount the events on the ground, to retell the many stories of trauma, there is a battle to be waged against prematurely fixing the image of pain and loss into static images and abstract accounts.

War is a contest between two sides, and injuring is its central activity. So might the building of Manhattan be described as a battle over terrain in which one victor remains standing after inflicting considerable injury. And memory is also a contest: it must be snatched from the hands of those who seek vindication and would narrate the story to legitimate their ends or rebuild the skyscrapers as a gesture of might. But memory must also be wrested from melancholy, which would turn the wound into sacred space in which loss and absence, mortality and death are contemplated without final acceptance of what has been lost. A site of mourning must acknowledge the singularity and irreplacibility of what has been erased. Yet it cannot retreat into private expressions that deny any awareness of shared experience. Because the shock of history turns catastrophic events into something intimate, it erases the distinction of what is private from what is public, forcing them to merge in unexpected ways. A public site of testimony and a private site of personal grief must struggle together over how best to memorialize the event. The site must be located in the present, but should look backward as much as forward.

This overwhelming trauma produced by the wound in the skyline and on the ground forces an exploration of two sets of graphic images and their related stories—those of the skyline of Manhattan and those of the collapsed World Trade Center. One narrates the glory of skyscrapers, while the other recounts the trauma of their demise. Implicitly, it is a struggle over the representation of the event of September 11, yet in looking backward in order to see forward, one must remember what an iconic image tries to summarize in its view. Ultimately, these memory-images—whether a distanced look at the skyline or the close-up view of the skyscrapers and their collapse—are surrogates for the human body whose actions created the skyscraper, built it from the ground upward towards the sky. The skyline is most powerfully a celebration of work and the competition that it entails. Horror and suffering lie in the opposite: in the unmaking of the artifact, in its ruin and wounds. In the skyscraper's demise, its image will henceforth be associated with expressions of death. Projected onto the human body, its memory-image is one of mournfulness and pain.

These acts of constructing and deconstructing are intimate and compassionate: they tie the skyline to the ground from which it materialized and to which it returned. They draw into a unity both creating and wounding, both the celebration of making and the injury of unmaking. While it may appear that jointly probing the two memory-images—a factual account of the capitalist conquest of the skyline and an emotional commentary on the horror and pain of September 11—is an impossible, contradictory exercise, it is the unity of the two that must be kept in balance if we are to mourn appropriately the memory of those who built and those who perished.

Building the Skyline

In the late 1880s, Manhattan real estate developers discovered that there was plenty of room to build into the air. By doubling the height of buildings, the same result could be achieved as if they had expanded the island to twice its present width. So with titanic force and energy, they began to run the buildings upward into the sky, until these very objects appeared to cry aloud of savage, unfettered strength, the outward expression of the freest, fiercest individualism. During this first round of construction in the 1890s, the architectural critic Montgomery Schuyler coined the term "skyline" to depict the immensely impressive image that "for a mile and more . . . reveals a chain of peaks rising above the horizon, itself a five or six story horizon, and struggling or shooting towards the sky. For another mile, for two miles more, the peaks continue to emerge, but they no longer form a chain." No other great city offered the voyager such a tremendously forceful sight—seen across a foreground of water there arose a jumbled mass of erections of various form and formlessness. It was a scene both confusing and impressive. Schuler lamented that the skyscraper was manifestly an "unneighborly" object, and that no building ever attempted to enhance the effect of any other. Consequently, the architectural excellence of the skyline resided in its parts, not the unattainable whole. Its resultant image was "not an architectural vision, but it does, most tremendously, 'look like business.'"

When Henry James returned for a visit to New York in 1905, he arrived by barge near the shabby and barely discernible rotunda of Castle Garden at the tip of Lower Manhattan. Yet he too was seized by the dauntless power of this most extravagant of cities. Its tall buildings, which now ran to fifty stories, were however, glory usurped—standing up like pins stuck anywhere and anyhow into an already overcrowded cushion. He found these skyscrapers to be "crowned . . . with no history, but with no credible possibility of time for history, and consecrated by no uses save the commercial at any cost, they are simply the most piercing notes in that concert of the expressively provisional into which your supreme sense of New York resolves itself." They did not speak of the long duration, of tower, temples, fortresses or palaces that are the built majesties of the world, but were instead the "monsters of the mere market."

During the great heroic period of skyscraper building, in the 1920s and 1930s, the silhouette of these enchanted mountains grew denser, and two distinct clusters of peaks began to compete for dominance: one in Lower and another in Midtown Manhattan. And while many praised them as dazzling shrines to a mechanized civilization, not everyone was enamored with the vainglorious outcroppings. Pondering on the fate of tall buildings in the aftermath of the disastrous earthquake in Tokyo in the early 1920s, Frank Lloyd Wright looked upon these "twisted, charred steel skeletons denuded of use that made them buildings, gutted from ground to top-rail. The thousands of human souls they housed were imperiled or have perished in the falling masses of masonry that had been pasted or tied to the steel skeletons in the name of 'Architecture.'" He called the skyscraper a commercial expedient that by its nature has compelled its neighbors to rise to similar heights, to compete or perish. Such an American get-rich-quick mentality was, moreover, "a dangerous expedient, dangerous to construct, dangerous to operate, dangerous to maintain and inevitably short-lived and destined to end finally in horrible disaster." No expedient is without flaw, Wright opined: "Millions of tons of brick and stone go high up into thin air by way of rivets driven into thin webs of perishable steel. Therefore millions of tons of stone and brick will have to come down again. Come down when?—Come down how?"

Yet build towards the sky they would. Hugh Ferris, the renderer of so many of these new peaks of the 1920s and 1930s, proclaimed to the *New York Times* in 1922: "[w]e are not contemplating the new architecture of a city, we are contemplating the new architecture of a civilization." In the

midst of the Depression, however, Wright continued to call each vainglorious skyscraper "an unethical monstrosity":

> It rises regardless of human life or human scale to impose exagger-
> ation on a weak animal. To keep it on him by persuading him it is
> his own "greatness," is no good. The herd instinct of the human
> animal is easy to exploit. The deserted agrarian areas of the United
> States testify to that. And this tall monument to the white-collarite
> is also testimony.

He added:

> I see it as monument to the same mistakes that spread unemploy-
> ment and disillusion from coast to coast and from Canadian bor-
> der to the Gulf. The one is out of the nature of the American
> people as much as the other. . . . Both issue from the abuse of
> privilege. Both bulk large and impressively as error often does and
> both are picturesque as error often is. War is picturesque.

Rockefeller Center, built in the 1930s, dealt the fatal blow to Lower
Manhattan in Midtown's competitive war. Conceiving it as "a city within
a city," John D. Rockefeller, Jr., removed eight decaying blocks of midtown townhouses to build a center that would accommodate close to
50,000 occupants and 16,000 visitors each day. Within ten years, it was a
combined business, international, and entertainment center with eleven
important structures erected on a site of 12 acres combined together in
superblocks. As Douglas Haskell praised it, although "[i]t was created in
the teeth of the biggest depression . . . it gave back to the city more space
and gave the people more art, and more joy, than any other 'city redevelopment' of near comparable size has done in two and a half decades of
trying with major help from agencies of government. Rockefeller Center
is the only large piece of urban renewal in business terms that the people
of the United States really love." However, Lewis Mumford vehemently
disagreed. In one of his "Skyline" articles appearing in the *New Yorker*,
"The mice have labored and they have brought forth the mountain," he

bitingly asserted. "Like most things that were conceived during the last days of the boom, the central building is very big. And when one has said this, one has said almost everything."

By the 1950s, the battle between the two skyscraper centers was clearly a thing of the past. Midtown Manhattan was the clear winner since it lay close to the hubs of transportation that tied the city to the suburbs, its skyscrapers were more abundant and modern, and it had residential and entertainment districts in close proximity. But Lower Manhattan was no sleeping giant to be brought to its knees without a good fight. It would wage war by bringing downtown some of the players who had built Rockefeller Center for yet another astounding assault and transformative redevelopment. This time the players would include David and Nelson Rockefeller, John D.'s scions, although the rules of the game had changed.

Every city planner learns that balanced community development depends on a healthy mix of commercial, industrial, residential, and recreational uses spread throughout the city and tied together with well-coordinated transportation infrastructure. Yet in post–World War II Manhattan, planners considered industrial uses to be obnoxious and signs of blight. They believed that Manhattan had too many small-scale operations in outmoded structures that by definition were inefficient, unproductive operations. Hence they set about trampling tens of thousands of blue-collar jobs under their urban-renewal bulldozers until they had completely eradicated manufacturing from the urban mix. Without any thought of the cost of creating a city whose core was middle class and postindustrial, New York would evolve into a global city reserved for white-collar services housed in gleaming skyscraper towers. If the skyscraper image is a celebration of white-collar production and the mighty forces of corporate capitalism, then it rests on a failure to celebrate the process of material production out of which it rose. Nor has it ever acknowledged the suffering it inflicted as it threatened the very existence of blue-collar workers. The competition ultimately split white-collar work and blue-collar labor apart, and only one would remain to claim Manhattan as its proper home. One of the epochal battles in this removal—notorious for all the wounds it imposed—took place on the blocks surrounding the site on which the Port Authority's World Trade Center would rise.

On the eve of World War II, New York was the nation's largest manufacturing city, with at least 40 percent of its workforce engaged in producing things. It was still robust at the beginning of the 1950s, even though Manhattan's priority as the central workshop was beginning to decline. Seeking ways to keep business from fleeing the city, Mayor Robert F. Wagner appointed a Business Advisory Council in 1954, cochaired by David Rockefeller and realtor Robert Dowling. These two men guaranteed that the city would create sufficient ground for thriving white-collar businesses and middle-class residential districts even if it meant destroying manufacturing areas. Since they clearly saw the emerging dawn of a service economy, they avidly supported the clearance of all old and deteriorating inner-city districts.

There were several areas of blue-collar employment along the fringes of Lower Manhattan. Two of these were market operations: the Washington wholesale vegetable and produce market on the West Side and the Fulton Fish market on the east. Just below Washington Market, on a site that would become the World Trade Center, stood "Radio Row," a consumer electronics district. David Rockefeller scorned these gritty businesses, viewing them as impediments to the financial district's expansion. While the hard core of the financial district reached skyward in a compacted area, he saw decay at its edges ripe for redevelopment. In 1956, he sponsored the creation of the Downtown–Lower Manhattan Association, which had a membership of some 225 businesses and financial interests. Not surprisingly, its report, issued two years later, identified nothing but obsolescence, deterioration, traffic congestion, and slow economic strangulation lurking in the financial district's waterfront shadows. Moreover, it argued that if the produce and fish markets were relocated to the outer boroughs and all unproductive land uses cleared away, then a golden opportunity existed to build in their place "walk-to-work" residential communities to service the financial core.

The logical path of expansion for the financial district, or so Rockefeller's billion-dollar redevelopment plan argued, lay to the northeast in a 564-acre swath of land along the East River waterfront, from Old Slip to the Brooklyn Bridge. In another plan issued in 1960, a three-story podium was to be located on a 13.5-acre site along the East River. On top of this plinth would rise a combination office and hotel structure of fifty to seventy stories, a six-story international trade mart and exhibition hall, and a central securities exchange building into which it was hoped the New York Stock Exchange would move.

David's brother Nelson was elected governor of New York in 1958, lending pivotal support to Lower Manhattan redevelopment schemes. But now the site for the proposed trade center began to shift from the east to the West Side, where it would completely dislodge Radio Row. By 1961 a deal was struck, enabling the Port Authority to take over the bankrupt New York–New Jersey Hudson Tubes and to include it within the new trade center. In addition, the World Trade Center was given an enhanced profile promoting Port of New York commerce by gathering together all businesses related to world trade. The Port Authority was now in full control of the biggest real estate venture to take place in Lower Manhattan, albeit by permanently eradicating much blue-collar employment.

The firm of Minoru Yamasaki and Associates was hired by the Port Authority in 1962 to develop architectural proposals. After a short visit to the site of Radio Row, Yamasaki comprehended the symbolic importance that Lower Manhattan gave to the World Trade Center. The site defied all his former notions of greatness, saying—or so he interpreted—look at us, we are not like others, we are full of energy, and we are willing to take risks. Symbolic of America, his edifice would lie within the gaze of the Statue of Liberty, a reminder of the country's self-image as an experiment in democracy. Standing above and apart from the huddled masses at its feet, the World Trade Center would add to the seismographic skyline of Manhattan the most outrageous peaks rising far above the debris of an outmoded port, a hemorrhaging manufacturing base, and disparaged blue-collar employment. The apotheosis of American-style finance capitalism, those towers were the heights of hubris.

As Yamasaki proclaimed, the World Trade Center "should not be an over-all form which melts into the multi-towered landscape of Lower Manhattan, but it should be unique, have excitement of its own, and yet be respectful to the general area." The Twin Towers were quickly nicknamed David and Nelson and, at 110 stories tall, they indeed stood apart from the mass of foothill skyscrapers huddled around them. Seeming to rise without end from a five-acre plaza, for a brief time the Twin Towers would be the world's tallest and biggest structures. Yet the towers never achieved the mythic status their builders desired, nor did they inspire "the loyalty and affection of New York's great skyscrapers of the past." As Michael Lewis wrote in their obituary in the *New York Times*, they made "a spectacular site but a lackluster performance, at best a colossal piece of minimalist sculpture—so ran the consensus."

Stratigraphies of Pain

On September 11, 2001, the devastating power of terrorism blasted apart the image of Lower Manhattan, leaving a tremendous void in the skyline of the city. What took seven years to build fell within a brief two hours. Long-standing debates about the merits and evils of skyscrapers seemed irrelevant in the face of such a catastrophe, which forced people to confront other, more pressing concerns. Almost instantly, the rubble of the World Trade Center was transformed into stratigraphies of pain.

It is too soon to know what iconic image will dominate our memory of the events of September 11 and fill in the gap in the skyline, although it must be an image strong enough to capture the enormity of the collapse of the towers and the severity of grief over the loss of life. An image is a vehicle to communicate a thought or a remembrance. It is typically backward-looking, reaching out to establish continuity with the past. And it should cause us to pause, to remember and think.

For some, the most evocative image may be that of Ground Zero itself, bringing to mind metaphors associated with the first atomic tests and all the paradoxes that entails. Ground zero is the center point of a nuclear explosion, where the ground melts away, where everything is swallowed up into a great nothingness. Hence the sign of zero, a circle drawn round a void, serves as the symbol for what was at the blast's center, for what was completely erased. But this is inevitably a problematic image, for a void at the heart of the problem raises issues of how to represent absence and how to think about the unthinkable. Diagrams of an atomic blast, moreover, portray rings of destruction unfolding from ground zero, yet these discrete rings camouflage the fact that the destruction in reality is a continuum, and while destruction may diminish with distance, there is neither containment within each circle nor closure of the contamination. Perhaps in this paradoxical manner, ground zero is an apt metaphor for the World Trade Center collapse, linking together problems of both the center and the periphery—the financial district and the rest of city, the nation, and the world.

No one individual seems to have named the site Ground Zero. Rather, the term appears to have been arrived at collectively. Given the level of destruction and loss of life, it is not surprising that so many immediately latched onto that powerful term. The site's rings of destruction encompass the nearby canyons of finance that stood silent and barren in the aftermath, as a numbing hush settled over the city. The name refers to the hundreds, then thousands who converged at the site, the "police officers, firefighters, part-time soldiers, bureaucrats, dog handlers, contractors, military veterans, staffs from emergency rooms, civilian survivors still coated with stinging ash, chaplains, massage artists, utility workers, misfits and thieves" who, according to the *New York Times*, "transformed the nightmarish bedlam into a new city within the old." And certainly, Ground Zero has come to represent the forest of cranes with their scissorlike hands crisscrossing the sky over the shattered amphitheater where the World Trade Center once stood.

The operation of clearing away 1.2 million tons of steel and concrete has a decidedly military quality. Divided into four quadrants with more than 150 pieces of heavy equipment, the debris in the beginning reached six stories high. Every night, attack plans were made with precision for each quadrant, for one false move of the heavy machinery, and the slurry wall—the concrete fortification—that keeps the Hudson River from flooding the site might be breached. Or empty pockets and holes within the 16-acre, seven-story basement underneath the World Trade Center site might suddenly open up, swallowing both people and machines.

Ground Zero, where the sinews of the city lay open and exposed, was a landscape like no other. Shards upon shards of steel lay twisted and intertwined; underground fires smoldered for weeks. Dangerous and emotional, it was far from being a normal construction site. As Daniel Hahn, a structural engineer at the site, explained: "when you watch TV you see a very antiseptic view of the W.T.C. collapse. . . . Down below you hear the cops shouting and getting excited. You hear construction workers talking construction worker language, you hear diesel engines of cranes. You have the aura, the smell and taste of death that's down there now. It's not a pleasant place to be."

For others, the iconic image of September 11 may lie in the images recycled again and again in ninety hours of nonstop television coverage: two commercial airliners flying into the quarter-mile-high Twin Towers, instantaneously causing both towers and planes to explode in an all-consuming fire fed by thousands of gallons of jet fuel. The terrorists were attuned to the fact that this stunning act of violence, carefully synchronized and ruthlessly planned, would be serially relayed as images around the world in quick succession. Theirs was a surrealistic gesture in which the touchstone of reality had to be reaffirmed—requiring Dan Rather to announce, "this is not a television graphic."

Maybe the iconic image lies in the plume of smoke that at first gushed upward into the narrow space between skyscrapers and then rained down upon the ground white ash, plaster dust, and paper debris. Barely discernible as the soft clouds of dust began to clear, ghostlike figures of ash-covered survivors struggled northward across the debris, their bodies wracked with sobs and pain, as they turned in horror to gaze at the site and grasp the unthinkable. One tower was no longer there, and then—horror upon horror—the other was gone.

The lasting image could be the heroic firefighters, policemen, and EMS crews who rushed to save as many of the victims as possible, the firefighters alone losing 350 brave men. Or is it the image of medics and doctors in their surgical outfits of green, standing outside Saint Vincent's Hospital silently waiting to attend the ambulances that never arrived? Or might the gripping image be the men who struggled to erect an American flag over the twisted steel girders rising above the smoldering debris?

And what do we do with the photographic image that caught a person falling head first after jumping from the north tower of the World Trade Center? Where do we put those scenes, incomprehensively described as "dark spots against the sides of the buildings" or mistakenly thought by a schoolchild to be "birds on fire"? Is it too soon to recall these traumatic and emotionally laden images of death? Can we remember the bodies that fell, imagine the pain and horror they suffered, or feel the grief of those who have nothing but photographs to take their place? Conceivably it was easier to grasp the simple "Portraits of Grief," the chronicles offered in memory of each individual victim that appeared daily for months in the *New York Times*. These images told tales about firefighters and stockbrokers, of dishwashers and fathers, of shoppers and brothers, of vice presidents and executive assistants, of so many hopes and lives that perished that day. And as they unfurled, they brought into focus a broader picture: the wide range of neighborhoods affected across the region, the disproportionate number of men and young who were lost, the cumulative deaths of 2,937 relatives, friends, and associates.

History shocks and arouses a sense of responsibility; big events become intimate in significant ways. As much as New Yorkers may mourn over the gap in the skyline and deal with the trauma of terror that brought the city to a state of emergency, we must remember that progress and catastrophe are inevitably linked; we must think through the power and hubris that put the image of the skyline in place. Its history involved destruction, pain, and loss as much as it did honor, glory, and might. It was erected by instruments that cut out and removed from the city uses and people considered to be inefficient, unsavory, and outmoded. Rebuilding the skyline of New York will be as much a battle over images and lost opportunities as it will be about material form. In this new struggle, we must stand ever vigilant, as Walter Benjamin warned, willing to wrest the image away from conformism that is about to overpower it. Just as we must always be aware that our need to memorialize may fetishize the act of remembrance and freeze it prematurely into static and concrete form. We need time to think through these images of Ground Zero, to remember that the skyline's lofty peaks derive from the strength of the lowly ground.

The Odor of Publicity

IN 1991, ON A SITE several blocks to the east of the World
Trade Center, archaeologists uncovered human remains in the
area now preserved and designated as the African Burial
Ground and the Commons Historic District. It is estimated
that the eighteenth-century cemetery contained the remains
of as many as 20,000 of the city's first African Americans, pre-
dominantly slaves who made up 40 percent of the original
Dutch colony and up to 20 percent of the English colony.
Though the burial ground had been very clearly marked on
old maps and surveys of the city, it had been built on succes-
sively since the late 1700s, and, at the time of the dig, was being excavated to make way for a new high-rise federal
building. When the bones came to light, embarrassing the
construction engineers, the cemetery was said to have been
"rediscovered." Disregard for its existence over the centuries
suggested all too clearly that the memory of this Colonial
slave population had become inconvenient to New Yorkers inclined to view their city as a bastion of the freedom-loving
North. Forensic examination of the remains showed the wear
and tear of severe physical hardship incurred through labor
and lack of nutrition, and during the slave ships' passage.
Many of the dead appear to have been literally "worked to
death." Not only had their resting place been forgotten, but
the marks of their suffering illustrated how they had been palpable victims of the earliest, belligerent efforts of the Dutch

West India Company to use Lower Manhattan to service international mercantile trade routes.

After the World Trade Center was laid low, it was possible to see the newly interred as victims of twentieth-century efforts to use the area in much the same way. After all, those who worked in the Twin Towers serviced a finance trading system that had taken its own, often brutal toll on populations around the world. In part because of the cruelty of that system's economic operations, these workers became the target of assassins who resented the power embodied in the highly symbolic buildings. Time will tell how their remains will be remembered, though it is most unlikely that they, or their memorial site, will be treated in the same way their African forerunners had been.

For one thing, the debate about the memorial and the replacement of the towers assumed a public character that has rarely been seen in New York City. The construction and the long-term economic impact of the towers had affected the lives of New Yorkers in very unequal ways—enriching some and impoverishing many others—but their painful collapse was embraced by virtually all citizens as a collective tragedy. In the matter of remembering and rebuilding, everyone could feel that his or her opinion mattered, even though few expected that theirs would count in any real way.

The impulse to review the past (with a view to building something better) threw up some instructive lessons. In my own case, it included the rediscovery of a prior site that I had missed several years earlier when I researched the World Trade Center history for a book chapter (in *The Chicago Gangster Theory of Life*) about the 1993 bombing.

The Twin Towers site marked the northern boundary of what had once been a Middle Eastern neighborhood called the Syrian Quarter, internationally known for its mercantile character. From the 1880s onward, it was the pioneer point of settlement for Arab immigrants in the United States. The term "Syrian" covered many countries of origin in addition to Syria: Lebanon, Palestine, Iraq, Egypt, and other Arab states. A bustling bazaar of a neighborhood, which became "the shopping center for those who sought the exotic East in the U.S.," it was shaped by community leaders into a model, immigrant settlement organized around "ethnic trade," and was much emulated in other parts of the country. The quarter also harbored some of the worst tenement housing conditions in the city, and it was squeezed by the expansion of the financial district in the 1920s and 1930s. In 1939, the authors of the *WPA Guide to New York*, who obviously had an eye for the exotic, noted of the neighborhood that: "although the fez has given way to the snap-brim, and the narghile has been abandoned for cigarettes, the coffee houses and the tobacco and confectionery shops of the Levantines still remain." Shish kebabs, knafie, baclawa, and other Middle Eastern specialities were widely available in neighborhood restaurants, and stores sold "graceful earthen water jars," "tables inlaid with mother of pearl," and "Syrian silks of rainbow hues."

It is sobering to recall that all of these commercial goods and practices, virtually continuous with the storied Levantine trade of the Ancient World, predated the Twin Towers' incarnation of a new kind of global trade—largely driven by financialization and the intangibles of stock valuation. World trade was also conspicuously present in the neighborhood's massive Washington Market, which overflowed with produce from every corner of the earth: "caviar from Siberia, Gorgonzola cheese from Italy, hams from Flanders, sardines from Norway, English partridge, native quail, squabs, wild ducks, and pheasant." When the long-disputed plans for the World Trade Center were implemented in the 1960s, the removal of the market was much lamented. So, too, the condemnation of Radio Row—almost fifteen city blocks of diverse retail and manufacturing in textiles, garments, electronics, and dry goods—provoked fierce resistance from hundred of those merchants who would be displaced. By that point, the building of the Brooklyn-Battery Tunnel had dispersed most of the Syrian Quarter to Brooklyn. Even so, there is no recollection of its existence in the primary literature on the building of World Trade Center, though its historical irony will escape few of us today, in a climate that has focused special attention on Arab-speaking immigrants. Like the African Burial Ground, its significance had to be rediscovered.

Aside from the power of nostalgia, which is not always justly applied, what can we learn from recalling the bygone profile of this neighborhood, formerly known as the Lower West Side (an appellation that has not survived to grace the lips of Manhattan's real estate brokers)? The scene I have been describing is marked by two urban forms that most of us think of as distinctive to New York City. Indeed, they are almost as distinctive, I would argue, as the city's skyscrapers have been, and they are just as important to honor in any plan for rebuilding on and around the site.

The first of these is the parochial neighborhood, sometimes classified as an ethnic neighborhood. The typical pattern of settlement is one in which a new immigrant group displaces an older one—in this case, the Arabs displaced Irish and Italians—and redefines the sense of the place with its imported customs, trade, and social patterns. It may be in large part due to the city's immigrant history that New York geography is still calibrated by its semiautonomous village communities and not by the numbered exits on its arterial highway system, as is the case in most American cities and suburban settlements. The social and economic tenacity of the neighborhood district has endured remarkably, despite the elimination of many such communities in the name of slum clearance and during the period of urban renewal.

The distinct microcultures of these New York City neighborhoods are not just the bread and butter of tourist guidebooks, or of real estate brokers for that matter. New Yorkers know that they are important markers of our essentially provincial identity within a city whose core is not at all unified or centralized. (Alone, for example, among the world's larger cities, New York hosts not one, but two central business districts, in Midtown and in Lower Manhattan, the second of which was kept alive artificially, as several urbanists have argued, by the World Trade Center plan to rescue and stimulate its commercial land value). In the weeks after 9/11, this *provincial* breakdown of the city by neighborhoods seldom seemed more pronounced. For Downtowners, our existing parochial maps of the city were reinforced by the drawing of frozen zones, security perimeters, military corridors, and other paralyzed pathways of transport, communications, and service.

The day after the attacks, I ventured down to Chambers Street, just north of Ground Zero. My neighborhood ID got me past the checkpoints, and I steeled myself for some surreal sights, like the crushed cars that had been piled on top of one another and dumped outside some of Tribeca's fanciest restaurants. Sleek emblems of Tribeca's boom years, these tony eateries were now service kitchens for relief workers, just as other neighborhood buildings would soon be relief centers for thousands of low-wage workers laid off by the collapse of small business and services around the World Trade Center. The *arriviste* neighborhood bourgeoisie were nowhere to be seen. The only residents on the streets were the artist-bohemian people who had moved there in the 1970 and 1980s, and had survived through two decades of ballooning rents. Just south of my building on Laight Street, I ran into people I hadn't seen in ten years. The hoarsening white ash thickened as I approached Chambers Street, buzzing now with new kinds of traffic. All of the insignia of authority— city, county, state, and federal—merged here, alongside fringe paramilitary organizations like the Salvation Army and the Guardian Angels (New York City's version of vigilantism, circa 1980). A machine gunner atop a jeep stood guard outside a shuttered bank.

I had lived in Tribeca for only eight years but had embraced the territorial posture that New Yorkers adopt to try and soften the edge of development in their corner of the city. In other regions of the world, neighbors talk about the weather. In this city, we talk incessantly about how our neighborhood is changing, and, in the grip of property capitalism's masticating jaws, it is usually always changing. In the forays I took in the weeks after 9/11, I found the impact of the tragedy had dramatized the role played by district boundaries. Old lines of demarcation were reinforced, like the checkpoints on 14th Street that barred outsiders from the frozen zone of Downtown. Even in the age of the three-thousand-dollar rental, Downtowners (below 14th Street) had still imagined themselves in that "other country" that wry nineteenth-century cartographers had depicted as *le pays de bohème*. Suddenly, now, this was the least urbane of urban quarters, and on our most civil of streets and avenues a freakish assortment of paramilitary vehicles (bearing unfamiliar acronyms) held sway. It was a shock to see the familiar jokes about the mayor's bunker mentality and the NYPD's army of occupation morph so quickly into the sinister reality of a neighborhood under official military quarantine. The New York that rose in the estimation of Americans in the wake of the attack was a city formed in the image of the WTC victims: corporate employees, public service workers, and in-person service workers. By contrast, the downtown geographies of ultraliberal New York, ultraethnic New York, countercultural New York, and queer New York seemed even more ghettoized than usual.

The second urban form to be noted is that compact mass of commerce and residence that is made possible by mixed-use zoning. In many city locations, the density and street activity in such districts approaches that of an open marketplace, the nearest thing we have, in this country at least, to the Middle Eastern souk. These are not simply zones of consumption. New kinds of business enterprise feed off the hum and scurry of their streetlife. Indeed, in the last decade, the kind of industrial growth that brought new jobs to the city—250,000 in the case of Silicon Alley's neobohemian digital industry before it faltered—occurred in the old manufacturing neighborhoods between Midtown and Wall Street which now host residences, retail, live/work, and nightlife. Despite the efforts by the Alliance for Downtown New York to attract full-time residents with mixed-use redevelopment, the business district still feels like a monocultural zone. The collapse of the World Trade Center may be the last gasp of the sequestered superblock, with its windblown corporate plazas, that fatefully eradicated the street. If the centralized banks and securities firms disperse in its wake, then urbanity will have registered a win in the long run. Small, shape-shifting firms more in tune with urban life and culture may well take their place, alongside much-needed housing stock.

Battery Park City (BPC), the umbilical soul mate of the Twin Towers, evolved as one version of an effort to resurrect traditional street patterns along with a mix of traditional housing forms—the townhouse and the low-rise apartment building. Despite its range of amenities—stores, cinema, restaurants, marinas, office blocks, esplanades—BPC has never shaken off its antiseptic profile as a security enclave where the public are temporary visitors and where the sharing of public space in its riverside park and piazza, while genuinely spirited, still feels like a privilege and not a right. This is a crucial distinction, since the issue of how and with whom we share space lies at the heart of our urbanity. Mixed-use zoning (with mixed-income provisions) is the only available technical instrument that can encourage, though never guarantee, that balance of residents, visitors, and strangers in free flow which gives urban space a sufficiently pub-

lic character. In the current climate of moral panic, we are more likely to
see the multiplication of those species of fear that thrive on the separation
of uses, a massive security retrofit that excludes all activities and persons
whom authorities and property owners might regard as "social anthrax."

In the immediate wake of 9/11, many of the city's signature architects—Johnson, Eisenman, Tschumi, Stern, Pelli—were calling for the rebuilding of the towers in some form, either to the same height or higher,
as a suitably august, symbolic response to the attacks. Community-
minded planners, along with the Regional Plan Association, were trying
to frame the rebuilding as an effort to encompass the metropolitan economy as a whole and stimulate the outer boroughs in particular through
better transportation links between the business core, its back offices, and
its newly decentralized operations. The gulf between these two perspectives was immense and telling. One proposal was vertical in its reach and
aimed at repairing the skyline along with the core economy of FIRE—finance, insurance, and real estate. The other was horizontal in scope and
took a stab at equalizing the region's uneven economy. Both involved
massive investments in physical infrastructure on a scale not seen or contemplated since the tail end of urban renewal.

This focus on physical design and planning was understandable. Architects and planners feel they have some professional control over physical detail, even if their plans are often overridden by higher ideologies.
(Look at Mohammed Atta and Osama bin Laden; one trained as an urban planner, the other as an engineer, yet each had an extracurricular
training in superideology.) The impulse to rebuild instantly captured the
public imagination as an opportunity to express the resolve of the nation.
Ground Zero, in other words, is already an ideologically charged site. Yet
its future and the future of public space in the city are surely entangled
with the fortunes of our own superideologies, especially privatization,
which has been number one for two decades now.

Thirty-five years ago, it was possible to build a development like the
World Trade Center only by using the legal instrument of a public corporation. Private enterprise could not have pulled off what the Port Authority was able to do using its supragovernmental powers, which are beyond
any public accountability; its public bond shelter from financial risk,

which no private developer could enjoy; and its capacity to exact tax
abatements from the city and state, instantly coveted and copied ever
since by corporations holding City Hall to ransom with the threat of
their self-removal to New Jersey. Who can compute what the balance of
tax revenue would have been if private investment had been left to pursue
its own course? Yet no one had their eyes on the prize of taxation when
the plans were conceived (by David) and approved (by Nelson) through
the kinship web of the Rockefeller family, or when they were executed by
the finance tycoons who sat on the board of the Port Authority. Nor
could tax revenue hold a candle to the treasure that lay in wait when the
World Trade Center became a pure instrument of real estate speculation
in Lower Manhattan. As a sweetheart deal, the whole thing was technically sweet—a developer's wet dream from first to last, consummated in
the name of a public that existed in name only.

The Twin Towers skidded into place just before the door closed on
large-scale urban renewal projects. Even so, the Port Authority's role in
the construction and management of the site was long subject to public
criticism, first from the left, and increasingly from the Giuliani administration, which saw the organization itself as a fat and sluggish holdout on
privatization's fast track. The sale of the Twin Towers' lease, just months
before 9/11, was a chronically overdue response to three decades of
heated condemnation. Yet would the public authority's sleight of hand
raise a hue and cry if it happened again? Public authorities, of course, are
still alive and kicking. Their most recent personification—the New York
State 42nd Street Development Project—brought us a $2.5 billion rein-
vention of Times Square that jet-propelled the "crossroads of the world"
towards Disneyfication as swiftly as the beachhead of the Twin Towers
guaranteed the financialization of the city's economy. Yet the primary
function of these public corporations—to assist land capital speculators
by assuming development risks that private enterprise itself will not
shoulder—is more and more likely to escape scrutiny. Why? Because, after two decades of steady privatization, it is routinely accepted that the
function of government is to assist the business of the private sector and
to accelerate, wherever possible, the handover of public assets to private
sponsorship. Where public corporations were originally a vehicle for New

Deal governments to get into the business of building public infrastructures, recently they have become an instrument of abdication, employed by governments to withdraw from all aspects of public life that show any indication of yielding market value.

In the wake of 9/11, and largely as a result of the courageous efforts of public service workers, popular faith in government has seen a rebound. The commonweal, we have been told, was back in the saddle, albeit at the price of patriotism. Even so, congressional Democrats exhausted much of their zeal by fighting for the mostly symbolic refederalization of the airport security workforce. With job losses soaring, and with the nation baying for decisive responses, there could hardly have been a better moment to justify public largesse and forswear private gain. Following its instincts, the Bush administration balanced the psychotherapy of emotional nationalism with its favorite kinds of neo-Keynesianism—bourgeois tax cuts, jumbo handouts to megacorporations in the form of alternative minimum tax backpayments, and lavish military contracts for Boeing, Raytheon, Lockheed Martin, and Northrop Grumman. The same formula is mooted for Ground Zero. Several groups and professional organizations (Regional Planning Association, Van Alen Institute, American Institute of Architects) offered plans and expertise for rebuilding, and the city's political heavies pushed for a broad, regional vision. Yet a large part of the mold has already been cast by Governor Pataki's establishment of the Lower Manhattan Development Corporation (a subsidiary of the Empire State Development Corporation) as the public agency which will oversee and fund all of the downtown rebuilding efforts. Its board, chaired by John Whitehead, a former chairman of Goldman Sachs, is top-heavy with business executives and political yes-men. Although Lower Manhattan boasts a diverse residential population of neighborhoods that stretch from Chinatown to BPC, only one residents' representative was named to the board of eleven, and she was herself the chair of the elite Community Board 1. The broad public is simply not represented in this lineup of decision-makers, who will have extraordinary powers to condemn land, apply funds, and override city land-use regulations in the name of the public good. Though modern political life is full of them, there can be few examples that illustrate more starkly the perversion of public planning processes.

The recessionary chill has barely abated the fever for financialization in the city that Wall Street likes to call the capital of capitalism. Those who hunger for a resurgent bull market are itching to re-create the kind of alpine built environment that embodied the brassy sway of a boomtown. Yet, in a district that has been plagued by hectares of empty office space ever since the opening of the towers, it would be sheer commercial folly to rebuild in the image of the high-altitude business god. Not so long ago, FIRE pundits were wryly recommending that portions of Lower Manhattan be allowed to revert to pasture in order to stabilize commercial property value. Besides, City Hall is already hooked on the habit of extending lavish tax incentives to developers for carving out residential space in the citadel of business. Corporate welfare notwithstanding, urban common sense dictates the restoration of the city's street grid to the 16-acre site, along with its conversion to mixed use and the provision of full pedestrian access from Battery Park's residential enclave to the place whence its soil came. As for a memorial, what better tribute than a new, genuinely mixed-income neighborhood that could capture, in the hustle and bustle of the living, the full sociological variety of those who died on 9/11? Against all odds, this kind of outcome might send a signal that the tides of privatization can be slowed, if not turned around, just as the building of the World Trade Center drove them on forty years before.

Letter to a G-Man

LET ME ASK YOU SOMETHING. Have you heard the story of the vizier's son? His father, the minister, had offended the ruler, and so he and his family were imprisoned for a very long time, so long in fact that the son knew only prison life. He reached the age of reason shortly after his release and, one night at dinner, the son asked his father about the meat he had been eating. "It's lamb," said the father. The son then asked the father, "What is lamb?" The father described the animal to the son, to which the son replied, "Do you mean it is like a rat?" "No!" said the father. "What have lambs to do with rats?" And the same continued then with cows and camels for, you see, the son had seen only rats in prison. He knew no other animal.

You may be wondering why I begin this brief correspondence with such a story, but I beg your indulgence. There will be time for all things. Suffice it to say that, as the son shows us, confinement defeats the imagination. Call it arrested development if you will, but if you are forced to stay put, how can you discover the delicacy of lamb, sprinkled generously with garlic and massaged with allspice, roasting over an open flame? Perhaps you can almost taste it now. Yes, the mind wanders, and the wanderer's mind, well, it expands, you could say. But without knowledge or history or experience, the son could only learn of these things when it was too late. I hope it is not too late for you—and for me.

You see, I fear that you have become like the son. You believe only what you already know, see only what you want to see, but you must ask yourself how you understand those things.

I have been told that you have arrested hundreds of us and seek to question thousands more. I imagine you are looking for me. You are concerned, naturally, after the eleventh of September, as we all are. I too watched the towers fall, as did everyone I know, with a tear in my eye and the air stuck hard in my lungs. Who could have imagined such malefaction! I prayed for the people lost in those towers, just as I have since prayed for the innocents everywhere, my benedictions sounding like Walt Whitman's brassy cornet and drums, which, as he said, play marches for conquer'd and slain persons. Didn't we all suffer on that terrible day, the families of the dead most of all?

The city itself was in mourning, with its gaping wound right there on the skin of Lower Manhattan. And here I am going to tell you something I presume you do not know. This is almost the exact same spot where, just over a century ago, the first of our extended Arab family came to this country. Have you ever wondered how Cedar Street got its name? I cannot tell you precisely, but I like to think it was because on Cedar Street, the Lebanese merchants from Zahle would sell you milk as sweet as honey and honey as rich as cream. We came first for the 1876 World's Fair, then began arriving in larger numbers, until in the 1890s we lived busily between Greenwich, Morris, Rector, and Washington streets. By the early part of the twentieth century, our community had expanded, reaching from Cedar Street on the north to Battery Place on the south. The western border was no less than West Street, and to the east, Trinity Place. But the center of our world was always Washington Street, a lane now blocked by emergency vehicles and ten-foot fences. To us, Washington Street was never just a street. It was our *Amrika*! After passing through Ellis Island, we would trudge up Manhattan Island with our weathered bags, looking for a friendly face in all the frenetic energy of New York, until we could hear a little Arabic and smell the food from home, knowing that on a street named for an American we had found Little Syria.

We came, like so many others, simply to make a better life for ourselves and our families. You could shovel gold on Washington Street, we were told, and so we trekked across the Atlantic, endured the verminous

Berenice Abbott. Lebanon Restaurant, 88 Washington Street between Rector and Morris Streets, August 12, 1936. Museum of the City of New York, Federal Arts Project

hostelries of Marseilles, and arrived with our satchels stuffed with hope.
City life was new to most of us, since we had lived typically in villages
and hamlets, and it was exciting. I remember what Abraham Ribhany
wrote back in 1914:

> New York is three cities on top of one another. The one city is in
> the air—in the elevated railway trains, which roar overhead like
> thunder, and in the amazingly lofty buildings, the windows of
> whose upper stories look to one on the ground only a little bigger
> than human eyes. I cannot think of those living so far away from
> the ground as being human beings; they seem to me more like the
> *jinnee*. The second city is on the ground where huge armies of men
> and women live and move and work. The third city is under-
> ground, where I find stores, dwellings, machine shops, and railroad
> trains. The inside of the earth here is alive with human beings; I
> hope they will go upward when they die.

His words never seemed so tragically real to me.

We came as sojourners, and after establishing ourselves in New York,
we launched out, men and women both, around the country as pack ped-
dlers. Loading up on goods from the stores on Washington Street, we carried what felt like the world on our backs. Our shops were fables to you.
Never had you seen our soft rugs for sale, or a gossamer web of silken lace
with Arabic letters hugging its border. Boxes rested on boxes in our tiny
dark shops, full of carved olivewood trinkets or luxurious satins or silver
wire as thin as a spider's web. As the *New York Tribune* put it in 1892: "In
the midst of all this riot of the beautiful and odd stands the dealer, the
natural gravity of his features relaxed into a smile of satisfaction at the
wonder and delight expressed by his American visitor. But the vision
ends, and with many parting 'salaams' one goes back to the dust and dirt,
the noise and bustle" of Washington Street.

We found no magic in our stores, however, just opportunity. We carefully folded the crocheted tablecloths of linen and stiff silk dress collars
and loaded them with the spicy perfumes and soft talcum powders into
our packs. The scrubbing soaps and gentle creams came next, and on top,

the rosaries, crosses, and carved icons that the people across this country so loved to buy from us, the Holy Land vendors. These are the things we carried. Jewelry and notions, we used to call them, and if you stopped to talk to us along our route, you might, as someone once said, buy a story with your bargain.

From the beginning then, our lives here have been about being on the move, carting goods and people across borders to make life a little bit better, a little bit easier, just a little more comfortable. We were the ones who brought the city to the country. We were Internet shopping before eBay, the catalogue before Sears. We went places others would not, namely, into the warm hearths of African-American homes, which ringed the cities we visited. There the food was heavier and the laughter heartier, and we would be treated to a hospitality we recognized like home. Detroit, Chicago, Fargo, Kansas City, Minneapolis, Fort Wayne, we knew the veinlike crisscrosses of this country before Jack Kerouac spoke his first French word. And we walked, mostly, and then we ached to come back to Washington Street, where we could replace our worn soles and enjoy a little backgammon before heading out again.

But that was a long time ago, and, well, nothing gold can stay. Maybe it is true that nostalgia makes time simple by the loss of detail, but today things seem so different. Since those early days, we have become doctors and lawyers, writers and engineers, but we are still shopkeepers and taxi drivers, and we continue to move lives around this country. And yet these days many of us sit stationary in our homes, unsure of what will happen to us if we step beyond the threshold of our doors. But I will come to that, all in good time, my good man.

We came from Mount Lebanon, from Syria and Palestine, but you called us all Syrians or, less accurately, Turks. We were mostly Melkite and Maronite, but there were a few Muslims, Druze, and Jews among us. By the 1920s, we had grown as a community into Brooklyn as well as Manhattan, on Joralemon Street, State Street, and Boerum Place, close to Atlantic Avenue, where you find many of our shops today. We continued to trade, and we worked in dusty factories, mostly sewing clothes and fine lace.

But in fact everything started to change in the 1920s. I talk not only about how, in the years leading up to that troubled decade, the immigration authorities became increasingly frustrated by our dusky looks, questioning whether we were "free white people" or "Asiatics." This racial Ping-Pong game used a strange chromatic logic that mostly bewildered us, and after the 1924 Johnson-Reed Quota Act and the harsh Depression of the 1930s, the numbers of our newcomers dwindled. Rather, I refer also to our daring to dream of self-determination back home.

After the door closed on the Sublime Porte, the lofty gate of Istanbul, the dissolution of the Ottoman empire was supposed to mean that we would have the right to determine our own fates. We thought you would support us, in the pioneer spirit of independence from foreign rule. But what we were left with were mandates and protectorates, leading to fracture and complaint in a moment when we felt unified and needed each other. The Europeans did not rule lightly, something I was sure you would have understood, but you have consistently lived up to underestimation, I dare say. It was the catastrophe of 1948, however, that broke our hearts. Tell me, what did the Palestinians do to warrant having their homes seized from them, their worlds disrupted, their lives bulldozed now for over fifty years? Because another people wanted the land the Palestinians had always lived on, they—the Palestinians—must be dispossessed into misery and squalor? Indeed the genocidal horror inflicted on the Jewish community in Europe was evil unmasked, but what had this to do with the Palestinians, except to turn them into the victims of another policy of extermination and cultural supremacy? It seems I am asking so many questions, but why you continue to deny the rights of the Palestinians just confounds me. It seems that their "crime" is simply to be born Palestinian, and in this scheme, a Palestinian life counts less than another. Yet there is no greater wrong in the world, for whoever degrades another degrades me and you and all of us.

Your ears prick up now that I am talking about the Palestinians. I think that when you hear this word, all you hear is terrorism. To us, we hear the echo of dispossession and the call for justice, but these days especially it appears to us that you are criminalizing all references to us and our Palestinian family, and it is affecting how we live here. For fifty years we have been speaking to you about this tragedy, but the actions of a handful of lunatics, madmen who have never until recently and only when convenient spoken about Palestine, have given you the motivation to shut us up and shut us down. You are infiltrating our mosques and gathering places, tapping our phones, detaining us by the hundreds, and seizing our charity. At airports you search us, and if you find Allah on a leaf of paper, you accuse us of sedition. We are beginning to wonder what you think you are protecting by all these actions—the people of this country or policies abroad that continue an injustice and lead to slaughter. But never mind that for now. There will be time. First, before you continue to cast us as perpetual foreigners, let me tell you why Muslim New York is our modern Granada.

For over half a century, we crossed the Atlantic to land on its avenue in Brooklyn. No doubt you know of this constellation of stores, restaurants, butchers, and bookshops, their wares piled high like the old stores on Washington Street. But does it surprise you to hear that our first recorded community organized around a mosque, back in 1907, stood not on this thoroughfare but in Williamsburg, Brooklyn, and was founded by a group of Polish, Lithuanian, and Russian Muslims? By 1931, this American Mohammedan Society had purchased three buildings on Powers Street for worship and community affairs. But Islam in this land surely precedes these intrepid travelers, for the first of us Muslims to arrive in this country dates back far before the birth of the republic. (You are confused because I had written we arrived in the late nineteenth century, and so you think I contradict myself. But I am large. I contain multitudes.)

Islam in this country is about as old as Virginia, and the first Muslims were brothers and sisters of our faith who were captured on the African continent and brought here solely for their labor. Have you read the slave statutes, like this early one, from 1670, which states that "negroes, moores, mollatoes and others borne of and in heathenish, idollatrous, pagan and mahometan parentage and country . . . may be purchased, procured, or otherwise obteigned as slaves"? We labored and suffered, and yet we continued to pray, fast, and recite the word of Allah whenever we could.

Take Ibrahim Abdur Rahman, for example. A son of royalty from Futa Jallon in West Africa, he was captured and made into a slave, land-ing in Natchez, Mississippi, in 1788. Over the next forty years, he was known to steal away to the riverbank when he could. There he would sit alone and scratch out Arabic words in the dirt and remember home. Later, the public learned about brother Ibrahim and his talents, and with his newfound notoriety, he sought to return to his people. Thus began a nationwide tour for Ibrahim. Paraded around the country by the American Colonization Society as an African curiosity, he raised money for his and his family's release from bondage and travel back to the African continent. This tour took Ibrahim not only to our New York but also to the White House, where he met John Quincy Adams. It seems the always polite Ibrahim had a sly, winking view of the politics of this country. He described his visit simply: "I found the President the best piece of furniture in the house," he states in a letter.

We are lucky to have Brother Ibrahim's story preserved. Most of our sisters and brothers who were enslaved have sadly fallen through history's sieve. We do have enough evidence, though, to know that Muslim slaves dot the forcefully tilled landscape of this country throughout its history and across its geography, from Natchez to New York and beyond.

In addition to this part of our family, there are the Muslim mariners, many of whom arrived in the ports of Brooklyn, ruddy-faced, out of breath, and eager for a place to bow their heads in remembrance of God. They surely came in the seventeenth, eighteenth, and nineteenth centuries. But we know that from 1939, after they landed they made their way to State Street, in the heart of the Arab community, where Sheikh Daoud Ahmed Faisal and his wife Khadija had their mosque, the Islamic Mission of America. (It is still there, but you must know that already.) In the cramped quarters of the brownstone mosque, sailor prayed with seamstress, African American shoulder to shoulder with Arab. It is said that the sheikh, by day employed by the railroad (again, on the road!), and his wife were individually responsible for spreading the faith to sixty thousand souls.

In fact, what we have always loved about this city is that we were never lost in it. By discovering each other, we found ourselves here. The Indian Muslims found the Albanians, the Malays prostrated beside the Africans, and all in front of Allah only. We didn't need mosques, only a clean place

to lay our foreheads gently on the ground. The sun gave us all the direction we needed. In those early years, like today, we converted brown-
stones and storefronts into prayer halls and mosques. And it continues.
Did you know, for example, that for the thousands of Muslims who
worked in the area around the World Trade Center there was a cavernous
room used for Friday prayer? From the beginning, we have lived here in a
kind of plurality that reminds me of Cordoba or Haroun el-Rashid's
Baghdad, and seems rivaled only by Mecca during Hajj.

But then after September 11 our halls and mosques had targets
painted on them, sometimes quite literally. What was for us a geography
of freedom and opportunity transformed overnight into a frightening
topography of rage. In the Bronx, our taxis were set on fire; in Manhattan, two drivers were beaten; in Bensonhurst, nine livery cars and taxis
were vandalized. Don't move, these thugs seemed to be telling us, because
we are coming for you. Death threats, physical assaults, verbal harassment, and a handful of murders across the country is what we (and our
brother Sikhs) endured. We were shocked and angry on September 11
too, and then we were afraid. When Timothy McVeigh bombed the
building in Oklahoma, was it right to seek retribution on any face that reminded you of him? (Instead, then too, we were blamed and we suffered.)
Vengeance is a strong emotion, but as Cleopatra tells her attendant
Charmian: "innocents 'scape not the thunderbolt."

By the smoke of my breath, we survived this terrible time with great
thanks to the grace of our neighbors. They deserve a thousand blessings
and one more, these decent, good-hearted people who wanted to help,
understand, and accompany us around our cities and neighborhoods.
They helped restore the streets as sites of circulation for us. But while all
this was happening, I daresay, now we have you to contend with. Do you
realize how you are chipping away at this sense of security we were just
beginning to feel again? I think you do.

There are many stories to tell, like our Afghan brother (shall we call
him Yousef K?) who was visiting his immigration lawyer's office in Lower
Manhattan and was stopped by the police. They inquired into his religion, and after he responded "Muslim," he was put into detention. Or
then there is the story of brother Butt. Someone must have been telling

lies about Muhammad Rafiq Butt, for without having done anything wrong he was arrested one fine morning. It was September 19, and the FBI was following lead 1556, a telephone tip from someone in South Ozone Park, Queens. The caller was concerned that two vans had stopped outside Mr. Butt's apartment building and six "Middle Eastern looking men" exited from each vehicle (no matter that Mr. Butt lived there with three other Pakistani men). After they arrested him, the FBI took a day to determine that this harmless 55-year-old man was innocent even to the temptations of the world ("He no smoke, he no drink, he don't go nowhere," is how his nephew put it). On October 23, after being detained for almost five weeks at the Hudson County Jail, Muhammad Rafiq Butt took his last breath and died that Tuesday morning, apparently of a heart attack. May God have mercy on his soul.

You see, my good man, we have lost our faith in your activities. You are turning what was for us an open geography into some kind of penal colony. Hundreds of us now languish in your prisons, not even sure why. You have admitted to the press that we have nothing to do with terrorism and that we have committed no crime, but still we cannot walk away, even if a judge has ordered us freed. Instead, you invoke an emergency, bond is laid aside, and we sit alone for 23 hours a day, the lights blazing the whole time so that night has lost its identity to day. Then you won't tell us who you have arrested. We have a difficult time finding out where our friends are as you fly them around the country with shackled legs and hands in midnight planes. You claim everyone has an attorney, but we have heard differently. You come in the middle of the night and take away our brothers and fathers and sons, and tell us nothing. Then you require us to "volunteer" for interviews, your reason for choosing us simply the kink of our hair, the caramel of our skin, the country name stamped on our passport. We have felt the freedom of the road in this country for a long time, and so you will understand if we are bewildered that this could happen here.

The other day, I heard a professor say that this was a time when we as a society should be thinking about what the balance between liberty and security should be, but the problem is that most of the country is willing to trade someone else's liberty—namely ours—for their own sense of security. He is a smart man, this professor, and he makes me wonder if this is the deal you have entered us into. While waiting for you, I have been reading James Madison. (Surprised? Didn't I tell you I have been here for over a century?) Since September, haven't we become vulnerable to the passions of the majority? I was under the impression that this required your greater vigilance for our safety, since, as Madison writes: "In a society under the forms of which the stronger faction can readily unite and oppress the weaker, anarchy may as truly be said to reign, as in a state of nature where the weaker individual is not secured against the violence of the stronger." You mouth the words of protection, but then why do we feel your violence lashing our backs?

Everywhere you say you are looking for rats, but I think you are finding lambs and unwilling to admit this. So many of us came here to escape terrible restrictions on our lives, not to rediscover them. But all around the world—in Chile, Iran, Iraq, Nicaragua, the Congo, Indonesia, Panama, and South Africa—hasn't the problem historically been not that you can't tell the difference between the rats and the lambs, but that you have preferred the rats?

Perhaps you would feel safer if I came to your office? Save you a trip? Under normal circumstances I would, but right now I would prefer not to. Like Bartleby, I have become a wanderer who refuses to budge. So send me off to the city's holding cells, the Tombs, if you wish. What will I discover there but the Egyptian masonry and forlorn history that lonely souls have scratched onto the stone in their spare time, for time is all they have in the Tombs.

In the meantime, they tell me that you are failing to fetch me, but keep encouraged. You may be missing me from one place, and so you search another. But I am here, my good man, under your boot soles. I am at home. I have stopped here, waiting for you. If I go anywhere these days, it is only to my roof, to hear the call to prayer from the mosque on Atlantic Avenue or the Sunday church bells on Pacific, and I sing along in what must sound like the yelp of a Barbary pirate to some. But to me these tunes are the sign of democracy. Don't you think so, too?

So come, ask me your questions. I will listen to them with devoted concentration, my head angled like a mendicant. But I won't answer them right away, for you must first have a sip of my syrupy coffee, a bite of crumbly sweet *halawa*, and a taste of our hospitality. There will be time for all things, believe me. And though you hardly know who I am or what I mean, I will be good to you nonetheless. We have much to discuss, you and I, and a long night ahead of us. *Yalla*, my good man, hurry and arrive. I've been expecting you.

From Jackson Heights to *Nuestra America*: 9/11 and Latino New York

ON THE BALMY, INDIAN SUMMER EVENING of September 19 in Queens, the Coalition of Colombian Organizations of New York has organized a candlelight vigil in solidarity with the victims of the World Trade Center (WTC) tragedy. It is 6:30 in the evening, and the Juan Manuel de Dios Unanue Plaza—located in the center of Jackson Heights on 83rd Street and Roosevelt Avenue—is filled with hundreds of Colombian immigrants and activists waving U.S. and Colombian flags in unison. Congressperson Joseph Crowley, New York City Council member John Sabini, Colombian Consul Mauricio Suarez Copete, and a wide array of local candidates running for public office all join the crowd. As the invited dignitaries mill about, Congressperson Crowley, a member of the House Committee on International Relations, briefs a small group of Colombian-American peace activists on the latest developments in U.S. foreign policy toward the civil war raging in Colombia. And as the evening progresses, the burgeoning crowd marches solemnly through Jackson Heights as they make their way to the local fire station. Respects are paid to the firefighters who lost their lives at the WTC, invited guests make short speeches condemning terrorist attacks in the United States as well as in Colombia, and candidates for local elected office strategically position themselves for the obligatory photo ops.

Clearly, the destruction of the WTC and the loss of human life was not merely a U.S. tragedy. New York is a global city with a large and growing multiethnic immigrant population, and not surprisingly, the 9/11 attacks claimed a cross section of the world's peoples. Moreover, the speed of modern communications highlighted the global dimensions of this unprecedented event in novel ways. Thanks to television satellite transmission, the residents of the South Bronx experienced the horrors of 9/11 at the same time as distant villagers living in Mexico, Ecuador, and other corners of the world. And to complicate matters, many of these same villagers had friends and kin who lived and worked in New York City and supported family members back home. So as the economy spiraled downward as a direct result of 9/11, negative financial impacts rippled outward from New York City and shook overseas economies that had grown increasingly dependent on immigrant remittances and the strength of the U.S. economy. The events of 9/11 in effect illustrate the collapsing of time and space and the weaving together of local and global events.

Immigrants stood out in sharp relief against the tragic backdrop of 9/11. Many worked in the economic sectors hardest hit by the attack's local fallout, and the responses of Colombian and Ecuadorian immigrants to it were particularly interesting. The distinct ethnic responses highlight how Spanish-speaking groups were similarly and differently affected. These distinctions are particularly important in a city that is increasingly marked by immigration and is rapidly assuming a distinctly Latino flavor. For example, although census data tend to undercount immigrants and other racialized minorities, the 2000 Census indicates that "Hispanics" account for 27 percent of all New Yorkers. Moreover, many of these "Newest New Yorkers" live and work in the multicultural neighborhoods of Queens—Jackson Heights, Corona, and Elmhurst. Colombian, Ecuadorian, Dominican, Mexican, and other immigrant groups have socially transformed these neighborhoods. They have stabilized them as well, buying homes and establishing small businesses en masse. Each group, by dint of hard work, has lived out in varying degrees the so-called immigrant dream. Yet they have responded to the WTC crisis in distinct ways.

Large numbers of Colombians have been migrating to New York City for close to forty years. Early arrivals were warmly welcomed as well-educated and disciplined workers. This changed during the 1980s, when Colombia emerged on the global stage as a major source and conduit for the distribution of illegal drugs. As a result, Colombian immigrants were often unfairly stereotyped as drug traffickers, and the welcome mat was abruptly withdrawn. And during the past two years, things became even more complicated as Colombia's political situation rapidly disintegrated. The long-standing bloody civil war, the use of terrorism as a generalized political tactic, the specter of U.S. military intervention, and the economic meltdown led to the migration of tens of thousands of educated middle-class Colombians to the New York tristate region. Many of these recently arrived Colombians, by dint of their undocumented immigrant status, found that their employment prospects were limited to poorly paid jobs in New York's burgeoning informal economy.

Colombians living in New York felt the effects of the terrorist attack acutely. On a raw and visceral level, the events of 9/11 brought to the forefront the global dimensions of political violence and terrorism. Because of Colombia's recent violent history, the realities of random political violence have a deep and unshakable resonance among many Colombian immigrants. Milena Gomez, a well-educated and articulate young woman, said: "I left my home in Colombia to escape the violence. Now it visits us in New York. In the modern world there is no place to hide from political terror." Diego Aguilera, a former professor of architecture in Colombia, added: "I feel deeply for what the United States is going through. Hopefully, the American public will now truly understand the terror that we experience every day in Colombia."

Terrorism is also an important political issue for Colombians living abroad. Immigrant peace activists have used the terrorist issue in lobbying U.S. politicians to support political amnesty for displaced Colombians living in the United States. In the wake of 9/11, the issue of political terrorism took on a new urgency among U.S. politicians. This political reality was not lost on a small group of Colombian activists in New York. The intention of the candlelight vigil in support of the victims of 9/11 was to demonstrate solidarity with the victims of terrorism in the United States and Colombia and to show local politicians that Colombian-Americans could successfully mobilize as an ethnic bloc. These goals had a certain political synergy. During this moment of national crisis and rising xenophobia, the vigil was a public and symbolic demonstration of immigrant allegiance to the United States. Moreover, the vigil also served to educate local U.S. elected officials on important Colombian political issues. It should be added that the use of public space and ethnic mobilization were loaded with political import. It said to one and all: we belong here, and as an organized bloc of Colombian-Americans we are a political force to be reckoned with. This was a political milestone for an immigrant group that had long been demonized and excluded from neighborhood civic and political spaces. And in light of the upcoming city elections, this manifestation of ethnic political mobilization took on an added significance. In short, this bottom-up example of immigrant political practice illustrates the fusion of local and global concerns.

The Colombian Consulate in New York City was, by and large, the institutional conduit for Colombian immigrants responding to 9/11. The consulate functioned as an information center and safe haven for many of the undocumented Colombian immigrants who were tragically touched by the events of 9/11. This was particularly important because many undocumented immigrants, fearing deportation, were reluctant to approach the local U.S. relief officials. Moreover, in their outreach efforts, consulate officials were very adept at commandeering and marshalling a wide range of modern communications technologies. The consulate's Web page posted the names of missing or dead Colombians, immigrant computer list servers efficiently distributed timely and pertinent information both locally and transnationally, and local Colombian radio stations assisted in grassroots mobilization and relief efforts. The Colombian Consulate also took the lead in organizing a memorial mass for the twenty Colombians reported to have died at the WTC. Colombians from across the greater New York region attended the mass at St. Patrick's Cathedral, seat of the city's archbishop.

At first light, it would appear that the popular and elite responses were complementary processes that reflected a common ethnic purpose. Yet Colombia is a violently contested terrain fraught with political tensions and long-standing contradictions, and the internal divisions there typically remain among Colombians who migrate elsewhere. The pervasive mistrust of the Colombian state and its political class is part of the baggage that immigrants bring with them to the United States. This distrust has largely marked the relations between grassroots Colombian immigrant organizations and the consulate. The efforts of the Colombian government to co-opt its overseas population as a subordinate player in nation-building and as a junior partner in refashioning its tarnished international legitimacy only reinforce these suspicions. Nevertheless, this was not the case with regard to 9/11. In large part, relations were surprisingly amiable and cooperative. When asked about this, Consul General Suarez Copete responded that as a rule "when there is a tragedy Colombians come together to help their brothers and sisters." This assertion rings true in this genuinely unique instance—the attack and its aftermath actually did facilitate smoother relations between the immigrant base and Colombian functionaries. Both sides limited their actions to mobilizing immigrant relief efforts, assisting their compatriots in securing relief assistance, and lauding ethnic solidarity in the face of political terrorism. And because U.S. officials and nonprofits crafted and controlled the rules and regulations that drove disaster relief, the Colombian actors were reduced to mere facilitators. This minimized tensions and disagreements.

Ecuadorian immigrants have been diligently establishing since the mid-1960s a social and economic foothold in New York City. Outmigration to New York accelerated during the 1990s, when the Ecuadorian economy and political landscape deteriorated significantly. The economic dislocations associated with the neoliberal trade regime, the official shift to the dollar, and increasing levels of political corruption set the context for mass migration. According to Hernán Holguín, the Ecuadorian Consul General in New York City, between 2 million and 2.5 million Ecuadorians have migrated abroad, with approximately 600,000 living in the New York tristate region. Considering that in 1999 Ecuador's total population was close to 12.5 million, this level of emigration is truly dramatic.

According to recent U.S. census data, 101,005 Ecuadorians[1] live in

---

[1] Immigrants and other subordinate populations tend to be historically undercounted by the U.S. census. Thus, the official U.S. census data on documented and undocumented Ecuadorian immigrants are highly questionable and significantly minimize the size of the New York City population.

New York City, with more than 50 percent in Queens County. Until recently, the growing Ecuadorian presence had largely gone unnoticed. Aída González, a well-known Ecuadorian activist, hinted at why this is: "Ecuadorian immigrants fall under the radar screen. We are stealth immigrants and don't attract undue attention. As immigrants our energies are primarily devoted to educating our children, making it economically in this country, and assisting our relatives back home." Moreover, unlike Colombian immigrants, whose relative notoriety derives largely from their demonization as drug traffickers, Ecuadorian migrants have not had to expend precious time, energy, and resources in justifying their presence in New York City.

While they are now a vital component in New York's rich ethnic mix, the fallout from the Trade Center assault also indicates just how enmeshed local Ecuadorians are in networks linking them to their country of birth. In a global economy characterized by increasingly permeable national borders, laborers, commodities, cultural products, and immigrant remittances flow quickly between New York and Ecuador. In many ways, these transnational flows are transforming space and place and converting many Ecuadorian households into appendages of the New York City economy.

In the year 2000, for example, $1,364 million dollars entered the Ecuadorian economy as immigrant remittances. According to the Ecuadorian Consulate in New York City, 40 percent of remittances are clustered in the country's southern region—the site of major emigration. As a result, remittances have altered local Ecuadorian household consumption levels, housing patterns, labor markets, and class dynamics. So when the economic impacts of 9/11 triggered unemployment in New York's highly internationalized immigrant labor markets—airports, hotels, restaurants, retail shops, and limousine services—the economic effects rippled through the Ecuadorian hinterland. As a worried cab driver observed: "The destruction of the WTC is financially hard on me and my family back in Ecuador. My weekly income went down by about thirty percent. What happens here in New York affects my family in Cuenca." In short, diminishing economic activity forced individuals and families into the uncomfortable position of having to tighten their belts in both

New York and Ecuador. These local-to-local, bottom-up impacts shed light on the indirect and global dimension of 9/11.

Coethnic immigrant organizations played an important role in structuring the grassroots responses to 9/11. For example, Ecuatorianos Residentes en el Exterior (ERE: Ecuadorian Residents Living Abroad) disseminated timely information to the families of victims in the New York Metropolitan Region. The Comité Cívico Ecuatoriano (CCE: The Ecuadorian Civic Committee) worked in tandem with ERE in organizing two requiem masses held in Corona, Queens, for the 27 Ecuadorian immigrants who died in the WTC. Internationally, the ERE leadership was also very active. They traveled to Ecuador with the explicit intent of assisting the relatives of WTC victims in securing humanitarian visas to enter the United States and claim the bodies of kin. The ERE leadership also, via the office of the Papal Nuncio in Ecuador, successfully secured a private audience in Rome with Pope John Paul II. At the meeting the ERE leaders broached with the pontiff their desire to have the Ecuadorian victims of the WTC attack named "as martyrs of terrorism and working heroes."

ERE's efforts in support of the Ecuadorian victims of the 9/11 attack were linked to a larger political project. Joseph Gavilanes, founder and president of ERE, is an activist in the grassroots and multiethnic immigrant amnesty movement in the United States. According to Gavilanes, the "undocumented Ecuadorian workers who died at the World Trade Center are victims of global exploitation." It is in this context that the symbolic shift from undocumented immigrants to working heroes assumes a political tint that is both local and global in hue. But ennobling working-class Ecuadorians both in New York and in Ecuador serves another purpose as well: to help in the struggle to secure immigrant amnesty in the United States and other nations.

Somewhat predictably, the Ecuadorian Consulate's efforts differed from popular immigrant responses. To a significant degree the Ecuadorian Consulate's efforts were very similar to those of the Colombian Consulate. In effect, the most important task was to coordinate the top-down dissemination of information. As a result, Hernán Holguín, the Ecuadorian consul general, spent a great deal of his official effort in preparing a

comprehensive list of the missing and dead and in coordinating the sharing of timely information in both New York and Ecuador. In rural
Ecuador it is difficult to disseminate information because of poor infrastructure and the geographic dispersal of many of the immigrant sending
communities. The radio station Radio Sucre helped circumvent this information bottleneck by communicating effectively with the many dispersed villages that dot the Ecuadorian countryside. Interestingly enough,
Delgado Travel, a firm established by an Ecuadorian immigrant in New
York, owns and operates Radio Sucre. The key role played by Delgado
Travel in the larger story of 9/11 is a vivid example of the dense transnational networks that connect New York City's immigrant neighborhoods
with Ecuador.

Like the attack itself, local immigrant responses to the horrific events
of 9/11 proved to be part of larger global processes involving transnational connections. In this sense, transnational networks sustain ethnic
identities in the face of assimilation and cultural homogenization. Yet
how are ethnic immigrant identities sustained in the face of an ongoing
external threat and a concerted global campaign by the United States
against "terrorism" and "terrorist sympathizers"? It would appear that in
the post-9/11 context, unambiguous immigrant allegiance to the United
States "trumped" any focus on local ethnic differences. According to
Monica Santana, the executive director of El Centro de Trabajadores Lati-
nos (Center for Latino Workers), "the crisis of 9/11 brought about a *tab-
ula rasa* that imposed a conformist kind of thinking that did away with
ethnic, political, and class differences among Latino immigrants."

This was clearly evident after 9/11 when Fernando Ferrer, the Puerto
Rican candidate for mayor of New York, distanced himself from his earlier campaign position championing the poor and dispossessed whom he
referred to euphemistically as the "other New York." This backpeddling
was clearly a strategic response to the political criticism that Ferrer was dividing New York City—partly by ethnicity, partly by class—during a moment of crisis. Moreover, many proclaimed that because unity was such a
priority, highlighting differences was unpatriotic. But shielding our eyes
from the New York City of the poor and the dispossessed simply rein-
forces the status quo of recent years and undermines the larger, overarch-

ing fight for social justice that concerns so many of the city's immigrants. It is a struggle that will resume, though—the divides in the city are simply too great to neglect. And despite the current bout of Americanization in New York, the ethnic diversity here demands that we continue to pay attention to ethnic differences.

Ethnic leveling, then, is by no means a predetermined outcome of globalization. The impacts and responses to 9/11 underline how immigrant activities are organized around ethnic anchors. The attack also demonstrated that different migrant networks follow different paths. Most important, however, the abrupt and dramatic transformation of the city lifted a curtain of silence and opened a window that permitted ethnic activists to make their presence felt concurrently in New York City and their respective countries of origin.

What Kind of Planning After September 11?
The Market, the Stakeholders, Consensus—or...?

WHO'S CALLING THE SHOTS on the rebuilding of Lower Manhattan—and how will the several billions in federal aid be spent that are targeted to New York City for that purpose? Money, after all, is going to be a key factor in every decision dealing with the consequences of September 11, in downtown Manhattan and elsewhere. On the face of it, the formal answer to the first question seems pretty clear: the Lower Manhattan Development Corporation, a subsidiary of the Empire State Development Corporation. Its formation was announced on November 6, 2001, just before the mayoralty election in New York City. Its members were named on November 29, 2001, just a month before the lame-duck mayor left office. It will have the extensive powers of eminent domain of its parent and the ability to override most local zoning and land-use regulations. Through it will be funneled the funds expected to be allocated to New York City reconstruction by the federal government, and the governor will likewise channel state aid through it. It will be a potent force, technically a decisive one.

Who is on it? Its members are summarized in a *Metropolis* article as one African American, no architects, no cultural leaders, one downtown resident, no educators, no families of 9/11 victims, three former Giuliani administration officials, one friend of George W. Bush, no planners, one union leader (construction), no urbanists, and four Wall Street executives.

They are, in the words of Herbert Muschamp, architecture critic for the *New York Times*, "captains of industry, including top executives from financial services and communications companies and from public agencies for construction and economic development." (Herbert Muschamp, "Leaping from One Void into Others," *New York Times* [December 23, 2001], sec. 2, pp. 1 and 42.) The chair of Community Board 1, the local planning body for the neighborhood in which the World Trade Center stood, is the local community's representative. The president of the Building and Construction Trades Council of Greater New York is labor's. The corporation is appointed by the governor (seven members, four-year terms) and the mayor, in this case the outgoing mayor (four members, initially one-year terms). The chair is Governor Pataki's designee, John C. Whitehead, former cochair and senior partner of Goldman Sachs. Whitehead served as the top deputy to former Secretary of State George Shultz and received the Presidential Citizens Medal from President Reagan. He was also chair of the Board of the Federal Reserve Bank of New York. Since 1999, he has donated $51,000 to the state Republican Committee and the maximum $30,700 to Governor Pataki.

It sounds more like the kind of heavyweight body that views itself as having unlimited authority to "get things done," in the style of the Robert Moses era, than a diverse, thoughtful, deliberating body guiding a city through a democratic process to elucidate its own general interest. That is not a reflection on its individual members; it is more a comment on the role its creators have established for it.

How will the corporation make its decisions? According to its chair, in his first press release: "The corporation will work closely with the private sector to determine a proper market-driven response to the economic and infrastructure needs of Lower Manhattan." Market-driven? Nothing about the World Trade Center and its site has been market-driven. Every development for the past forty years—at least—has been done through the good offices of the state, in fact quite counter to the market. The private sector may indeed have been behind what was done, but not operating through the market. The World Trade Center site was condemned by the state, and some 800 small businesses that were on it (along with perhaps 30,000 jobs) were displaced through the power of eminent domain. The city closed streets in order to create a new superblock, and the planning and design of the World Trade Center were paid for by the state. After it was built, there was so little market demand for space in it that the state had either to lease back the space to itself or leave it vacant. And the whole concept of the trade center in a financial district location was originally promoted by David Rockefeller, who was anxious to shore up the value of Chase Manhattan's real estate investment in the area.

The decision came at a time when the market had been settling on Midtown, rather than the financial district, as its place for expansion. Indeed, the market still favors Midtown today; the notion that New York City's office sector is dependent on its location in Lower Manhattan is a myth. Bob Fitch's excellent *Assassination of New York* is essential reading for the story; Eric Darton's *Divided We Stand* provides useful detail. While public investment in the World Trade Center did indeed stimulate a spate of private investment in the area, there are today 182 million square feet of office space in Midtown and 65 million in Midtown South, but before September 11 there were only 82 million square feet in Manhattan below Canal Street. The Midtown area had grown much more rapidly than the Downtown, and in Downtown before September 11 the city was being asked to invest heavily in tax and other incentives to pay for the transformation of some of its office space to residential uses.[1] Today, estimates are that at most 2 million square feet of office space would be needed in new construction to make up for the 15 million square feet that disappeared in the attack and the more than 10 million feet rendered at least temporarily unoccupiable.

In fact, the movement of the market is not towards the rebuilding of the World Trade Center or its equivalent in office space, but rather away from Lower Manhattan altogether. Under pressure of immediate necessity, occupants of the affected area took space wherever they could find it

---

[1] Figures are from page 1, Real Estate Section, the *New York Times* (January 6, 2002). Midtown is defined as between 65th and 34th Streets, Midtown South between 34th and Canal, and Downtown below Canal Street, excluding Tribeca.

elsewhere. But many of those who moved entered into long-term leases, built out offices they already had elsewhere, and saw their moves as part of a longer-term strategy to deconcentrate from Lower Manhattan rather than put all of their eggs in one basket. The market is moving across one river towards Jersey City and the edge cities of New Jersey, and across the other river to downtown Brooklyn, where the office complex known as Metrotech is burgeoning. The roster of companies taking large blocks of space on long-term leases outside Lower Manhattan includes Citigroup, Merrill Lynch, American Express, Lehman Brothers, and MetLife. It's a pattern that had been going on for some time before September 11 and which has accelerated since. Moreover, the angst resulting from the collapse of the towers will last for a long time. Even Larry Silverstein, the owner of No. 7 World Trade Center and the 99-year leaseholder for the Twin Towers, who, of all people, has the incentive to sound upbeat about market demand for his space, says frankly that it will rent well because he expects to be able to offer below-market rents to tenants.

Whether the market will even promote a continuing growth and high-density clustering of financial services firms anywhere, let alone Lower Manhattan, is itself an open question. Chase Manhattan gambled that it would, and built in Lower Manhattan in 1956. Citibank, the largest U.S. banking firm, gambled (or invested) differently, and built in east Midtown. With the end of what had looked to optimists like a permanent boom in the stock market, some of the luster has gone out of businesses whose sole activity is trading or speculating. And once the implications of the Enron collapse are digested, there will be even less enthusiasm. The state has viewed finance in recent times as the engine of New York City's economy and thus favored it in many ways, including land-use regulation. But the market alone has not accounted for New York City's growth; other factors were also at work. Now, in fact, the weakness of the market may contribute to a slow shrinking of the city.

A "market-driven response" is hardly what those most concerned with the details of rebuilding in Lower Manhattan have in mind. The public investment in infrastructure, particularly transportation facilities, that would provide a solid basis for such a market needs to be immense. But nobody is pricing out the facilities being talked about: a brand-new

PATH station to let commuters from New Jersey travel conveniently to the area; linkage to a brand-new and complex giant subway and commuter hub around the existing Fulton Street complex; a Second Avenue subway; waterfront improvement, cultural facilities, subsidized housing, tax abatements—all this would easily mount into the tens of billions of dollars. Whatever drives such an investment is hardly the market; the market might very logically call for transportation investments in the hubs already existing at Pennsylvania Station, Times Square, and Grand Central, instead of creating a brand-new one in the financial district. Real estate investors and financial firms with a vested interest in Lower Manhattan might indeed now push for public investment there (as they did when the World Trade Center was built), but other market interests, still somewhat quiescent, will clearly push in a different direction. Seymour Durst and other Midtown real estate owners have already begun voicing their sentiments, and one can expect their influence to be felt more and more as the magnitude of the public expenditures necessary to support "market-driven" investment in Lower Manhattan becomes apparent. Market players also respond when some of them are taxed to support others, after all.

Indeed, the "private sector" has been heavily involved in everything that went on with the construction of the World Trade Center and is heavily involved with everything that is going on with the site now, but it is hardly working through the market. When it is in its interests to do so, the private sector operates in the public sphere, market be damned, and that is what is in the offing here.

However, in addition to looking for a "market-driven response" and working with the private sector, the Lower Manhattan Development Corporation wants to take a broader look: "In addition to the Board of Directors, the corporation will also establish a broad-based Advisory Council. The Council will seek input from a wide array of community groups, business owners, elected officials, other interested parties in the redevelopment of Lower Manhattan. The Council will make periodic recommendations to the Board." This is the so-called stakeholder approach to planning, very fashionable these days, and the chair of the corporation indeed speaks of "stakeholders." Who does he suggest they are? "[T]he families of victims, downtown residents, commuters, Wall Street firms, developers, retail shopkeepers, and cultural organizations." He speaks of naming committees from "various constituencies," but what are they? "Real-estate executives, downtown residents, bankers and victims' families."

Hardly an exhaustive catalogue. Of those who lost their jobs as a direct result of September 11, less than 20 percent lived in Manhattan, and probably only a small fraction of those were from Lower Manhattan. Labor unions represented many of the victims; immigrant and ethnic organizations did as well. They should all be at the table. The Federal Emergency Management Agency (FEMA) has defined the entire city as an area where renters or owners are eligible for special disaster assistance on rents or housing costs if the problem was caused by September 11, but very few of them will be in Lower Manhattan. The other boroughs have long had a vital interest in what was and was not done in Manhattan; they ought to be centrally involved now. Transportation used to be seen, legislatively, as a tristate concern when we had a Tristate Planning Agency; all three states have stakes in transportation changes in Manhattan. It is good that one community board is represented on the corporation, but the adjoining boards, and boards around the city, all have stakes in what happens. One report, for instance, discusses the need for investment in Brooklyn and Queens. Such investment raises complex issues, and other boards left out of the process so far may also want to participate.

The brouhaha over the visitors' platforms is one example of what's wrong with the narrow "stakeholders" approach. Not surprisingly, several survivors objected, with one calling the visitors' platforms "ghoulish." Interestingly, though, a teacher from Hood River, Oregon, disagreed with this sentiment in a letter to the *New York Times*. She had taken a group from Hood River to the viewing platform "not [as] a grotesque display of curiosity, but [out of] a desire to become more a part of what you had had to deal with. . . ." Another person whose fireman brother died on September 11 wrote: "it's clear that this event has had a significant impact on many people who do not live in New York." Also, a writer from Michigan reported: "I hope to be there soon [with a] need to get close enough to feel the pain and be able to share the view with my kids, grandkids, and friends. . . ." Our leading politicians have been stressing that the attack on the World Trade Center was an attack on "all Americans." Are they not all "stakeholders" in what happens there, whether they live or own property there or not? Simply put, setting up committees based on "stakeholder" representation does not work as a planning process.

There is an alternative way of making decisions about Lower Manhattan. Make the process transparent, democratic, and informed. Doing that is not easy, but it involves two things: a maximum of public information and discussion, and a maximum use of the democratic planning and budgeting procedures that were long fought for and are now enshrined in law.

As for information and discussion, several civic groups have already begun to render yeoman service. A special task force, under the sponsorship of the New York City Partnership and seven consulting companies, has produced massive documentation on the exact economic impact of September 11. Although the partnership represents a clearly defined and limited range of interests (the "voice of New York City's business community," as it describes itself), the information it provided is valuable. Both the New York City Partnership and New York New Visions, consisting of some 20 professional organizations in planning and architecture, under the aegis of the American Institute of Architects, issued strong recommendations on rebuilding plans. Somewhat narrowly focused on Lower Manhattan, R-DOT (Rebuild Downtown Our Town) is an active group based in Lower Manhattan (broadly conceived) that advocates vigorous public participation in the process. Mobilization for Youth has helped pull together a coalition of grassroots groups that hope to exercise influence. The Civic Alliance, assembled by the Regional Plan Association, is perhaps the broadest of the civic amalgamations, including now some hundred of such groups in its coalition. Other grassroots, community, and labor groups are also active. The Community Labor Advocacy Network is what its name implies, and it also seeks influence on what happens. The process needs discussion and critique, and these groups are all doing what they can to forward public debate.

Some of the civic groups, however, have in mind a "consensus" image of planning. In hoping to minimize conflict, they ignore its real existence. This is a danger for professionals in particular, who sometimes act as if

their expertise will enable them to formulate conclusions about the public
interest as if there were no conflicting interests involved. And there is a
potential further problem as Herbert Muschamp says:

> The aggressive response of corporate architecture firms and their
> developer clients to the terrorist attacks was not New York archi-
> tecture's finest hour. Nonetheless, much of this behavior can be di-
> agnosed as a symptom of the privatization of city planning over the
> last two decades. Never has this surrender of civic responsibility
> been more starkly illustrated than in recent weeks. . . . Corporate
> architects, despite clear conflicts of interest, try to act like public
> servants, with evident sincerity in a few cases.

The codes of ethics of two of the leading organizations among the
civic groups in the discussions, the American Institute of Architects (AIA)
and the American Planning Association (APA), both speak of "serving the
public interest." The AIA code, for instance, holds: "members making
public statements on architectural issues shall disclose . . . when they
have an economic interest in the issue" (Rule 2.301). The APA Ethical
Principles in Planning code holds: "The planning process exists to serve
the public interest." It goes on to say that participants should "make public disclosure of all 'personal interests' they may have regarding any decision to be made in the planning process in which they serve, or are
requested to serve, as advisor or decision maker" (Section B.2 and B.3).
The organization defines "personal interest" broadly to include any actual
or potential benefits that they, a spouse, family member, or person living
in their household might directly or indirectly obtain from a planning decision. Yet so far, no professional participating in the discussions has revealed any interest in obtaining commissions in Lower Manhattan or
work for any client that has an interest in developments there. The point
here is not to call into question the ethics of any individual professional
in the field. It is rather to highlight the implicit assumptions that professionals often hold: that professional expertise can resolve what is best for
the public; that professionals can arrive at conclusions uninfluenced by
their own interests or those of their clients; and that real conflicts of inter-

est do not exist or are only matters of inadequate technical understanding which professionals should try to convert to consensus.

Despite all the professional actions, the good intentions, and the diverse ideas for rebuilding, money is the basic force that will determine the outcome. And unfortunately, naïveté about money is the largest hurdle facing many of the civic groups pressing for greater participation. The public discussion about money has hardly started, and the civic groups are thus far not visibly engaged. But how, by whom, and for whom public funds are allocated will be critical in the decision-making process. Real estate interests have been active ever since September 11, lobbying city officials, the state, and New York's two senators about how much money is needed and for what. Robert Kolker, in *New York* magazine, describes one such important meeting:

> "Get the money now," Bill Clinton is chanting into a microphone, sermonizing before a parish of pinstriped suits. "Get the money now." [The meeting] is in Citigroup's packed Park Avenue auditorium with some of the city's top business leaders—megadeveloper Jerry Speyer, billionaire financier Henry Kravis, real estate baron and publisher Mort Zuckerman, AT&T chairman Mike Armstrong, Lehman Brothers chairman Dick Fuld.... Looking on from the audience like proud parents are Hillary Clinton and Chuck Schumer, along with George Pataki's economic czar, Charles Gargano, and union chieftains Brian McLaughlin and Randi Weingarten. "Get the money now," Clinton repeats in his folksiest drawl. The terrorists aimed for the World Trade Center because they "think we're weak and selfish and greedy," he says, but that's no reason not to fight hard for that money.

Civic groups, of course, are not quite as experienced at chasing (and getting) the money as our ex-president and don't have quite the influence of those Clinton was addressing. But that's where the focus needs to be if democratic planning is really what we're after.

Spaces of Reflection, Recovery, and Resistance: Reimagining the Postindustrial Plaza

A NEW DEFENSIVENESS HAS ARISEN since the September 11 terrorist attack. Concrete barriers, private guards, and police protect what were previously open plazas and buildings. The threat comes not only from the outside—from fanatics who hijack airplanes and crash them into buildings or send anthrax through the mail—but from the danger that Americans will overreact to the destruction of the Twin Towers by barricading public spaces and denying ourselves opportunities for expressing community, openness, and optimism.

Long before the World Trade Center bombings, the public spaces of the postindustrial American city had been fenced off, policed, and privatized. Moreover, antiurban sentiment and fear of the "other" across social classes has been a mainstay of residential and workplace segregation ever since the development of suburbs. People began moving to them to escape the dirt, disease, and immigrant populations of the inner city as soon as trolleys made it feasible. And suburbs offered more than just a physical distance from the city—a more powerful social distance emerged, maintained through a complex social discourse of racial stereotypes and class bias. But even in cities, similar forms of social distance took shape. Today, for instance, wealthy New Yorkers deal with this fear by living in separate zones and limited-access cooperative apartment buildings. Other city residents rely on neighborhood watch programs and tolerate increasing restrictions on behavior in public spaces.

Typically, Americans have translated this more generalized culture of antiurban sentiment into a more specific fear of the crime and violence they say pervade the city. And even in the face of declining crime rates, the fear has ended up justifying more rigid controls of urban space, particularly in socially diverse public spaces.

Indeed, many of these civic spaces are no longer democratic places where all people are embraced and tolerated, but instead centers of commerce and consumption. Collaborative public/private partnerships between municipalities and local businesses have transformed such places as Bryant Park in Midtown Manhattan into safe, middle-class environments maintained by surveillance and police control. Similarly, the public space of Battery Park City—built on a landfill created by the World Trade Center's construction—has limited public access and a design vocabulary that appeals mainly to the upper-middle class.

Unless North American urban spaces become commercially successful, their future remains in question. Commercialization and privatization, however, limit participation to those who can afford it and conform to middle-class rules of appearance and conduct. The mall-like interior of the World Trade Center was one of these spaces of consumption, bringing people together through transportation nodes, restaurants, and shopping (ironically, the World Trade Center's shopping mall was probably its most genuinely public space). Interior public spaces such as the Winter Garden at Battery Park City also possess these qualities of commercialization in addition to an unambiguous police presence, but at the cost of being separated from the greater city they should connect with.

These are not the places to heal and reconcile our hopes and dreams. Instead, it is the more spontaneous, less-regulated spaces to which we should turn. Union Square, with its green trees, benches, and circular walkways, became a sacred space through spontaneous memorials of written messages, flowers, candles, photographs, and other objects commemorating the victims of 9/11. The construction fence surrounding Ground Zero was also decorated with messages, prayers, and personal memorabilia. Other places of grief are models as well. Mourners placed handwritten prayers, children's drawings, poems, teddy bears, and flowers on the chain-link fence around the Murrah Federal Building in Oklahoma City, site of the 1995 Oklahoma City bombing. And in a park across the street from Columbine High School, mourners created a shrine the size of a football field with messages, paper chains, and crosses to remember the twelve students and one teacher who were shot in 1999.

Other nations provide models of democratic, open public spaces as well. For example, the Plaza de Mayo, the central public space in Buenos Aires, Argentina, evokes the grief and political resistance of the mothers of the disappeared. In the center of Mexico City, the Zocalo commemorates the Spanish conquest and colonial history, and at the same time represents past and present struggles for indigenous representation and citizenship.

Given these models, the World Trade Center site provides an opportunity to reimagine the postindustrial plaza as a space of reflection and recovery as well as a place of civic action and discussion rather than a privatized space driven by global capital. This site of trauma can be transformed into a communal center for people to meet, mix, mourn, and remember.

The "publicness" of spaces consists of five kinds of spatial rights: access (the right to enter and remain in a public space), freedom of action (the ability to carry on activities in the public space), claim (the ability to take over the space and resources in it), change (the ability to modify the environment), and ownership (the ultimate form of control). Yet fear now pervades the postindustrial plazas of New York City, and these rights have been curtailed by surveillance and policing with guard dogs, inhumane bench and ledge design, and the more subtle cues of high prices and luxury boutiques. In addition, barricades for special events and seemingly permanent scaffolding for construction often make accessing the sites a daunting task.

The enhanced fear of terrorism—evidenced by increasingly novel surveillance techniques—is making things worse. New electronic monitoring tactics are being implemented across the United States. Before September 11, the idea that Americans would agree to live their lives under the gaze of surveillance cameras or monitored by police seemed unlikely, except in the privatized public spaces mentioned. But now citizens are asking for outdoor cameras to be installed on the boardwalk of Virginia Beach to scan faces of people at random, cross-checking them with faces of criminals stored in a computer database. Palm Springs is wiring palm trees to put electronic eyes on the main business street. What was once considered "Big Brother" technology and an infringement of civil rights is now treated as a necessary safety tool, without an examination of the consequences. What is at stake is the cost we pay for fighting terrorism, measured not just in salaries of increasing numbers of police officers or eye-recognition technologies, but also in the loss of freedom of movement so characteristic of the American way of life.

The current obsession with monitoring public life, then, portends a narrow space of remembrance at Ground Zero, one that can be easily surveyed and controlled. This raises a host of questions. Since sites of trauma are about the production of meaning, whose stories will be communicated and interpreted? A monument organizes historical memory, but whose history will be recorded, and towards what end? A site of trauma embodies the dead buried there, yet the interpretive story transcends the actual site, capturing the diversity of all New Yorkers affected by the attack. Many audiences—downtown residents, victims' families, firefighters, volunteers, city institutions, displaced companies, to name a few—have a vested interest in being represented. More marginal groups also have voices in the debate, which further complicates the situation and increases the possibility of disagreements about what should be expressed.

The diversity of responses to the attack, then, suggests that any single, homogeneous design for the memorial—especially one that follows the current model of policed space—is not what is called for. Rather, the site must respond to the different experiences and reactions of people throughout the city, divided as they are by age, generation, location, ethnicity, and class.

New Yorkers have responded differently to the tragedy on the basis of their previous experiences of coping with crisis. Previous life events provide a context and offer strategies for response. For example, people in their eighties who lived through World War II and remember the Holocaust are not as shaken by the terrorist bombings. They have experienced death and tragedy before, and tend to view it as an inevitable part of everyday life. Americans in their forties, fifties, and sixties remember Vietnam and the Cuban Missile Crisis. I remember endless school drills, crawling under my desk or crouching in the hall, practicing what to do in the case of nuclear attack. My neighbors built elaborate bomb shelters to house their entire family. Terrorism is something I have lived with during fieldwork in Central America and Africa, so the September 11 attacks were not a shock. Young adults in their twenties seem most affected by their shattered expectations and sense of the future. They have not experienced crises of this magnitude before, and their needs and concerns are different from their parents' and grandparents'.

People also responded to the terrorist attacks differently on the basis of where they live and work in the city. Individuals' mental maps of the city—personal impressions and physical relationships that order a person's orientation in space—were destroyed, and their sense of place attachment—how they feel about the spaces they live and work in—was disrupted. How close you live to Ground Zero seems to determine the degree of cognitive disorder. It ranges from mild disturbance in day-to-day transit about the city to abject mental anguish and anxiety. Those residing nearby have experienced the latter. A young man who lives nearby in Battery Park City finds his world turned upside down—he misses his friends, his routines, his ways of moving across the city, and his favorite places of relaxation. Moving uptown has only further dislocated him, and he hopes to return downtown soon. A female friend who lives near the site in the din of trucks, night lights, and volunteers complains that she can no longer sleep at night, find a grocery store, or orient herself when she goes outside.

Anthony Oliver-Smith, an anthropologist working in the very different context of Peru, observed that when a massive earthquake and avalanche destroyed the small city of Yungay in 1970, survivors reacted with disbelief, wandering aimlessly looking for remnants of their homes. The tops of four palm trees marking the original plaza were all that remained of the original site, and they became the centerpiece of a new plaza and town, rebuilt against the advice of experts, on the same site. Interestingly, though, he noted that *mestizos* and indigenous Peruvians, between whom lay a stark class and ethnic divide, responded to the tragedy in contrasting ways. For instance, the Yungay disaster gave greater power and social control to indigenous groups, because so many members of the *mestizo* middle and upper class had been killed in the tragedy. The middle-class survivors had to adjust to their altered social position and identity in addition to all the other problems and concerns they faced.

Of course, a shift of this magnitude could not occur in a city the size of New York. In fact, the large number of residents affected makes it difficult to estimate the degree to which the terrorist attacks changed people's lives. But what we can safely take away from this other tragedy is that there are important differences in the responses of culturally identified individuals in the communities closest to the World Trade Center—Chinatown, Little Italy, the Vietnamese community, the remaining Jews on the Lower East Side, and the actively growing Latino enclave living nearby. There are also significant differences in effects. Essentially, New Yorkers are not responding to this crisis in similar ways or in ways that any one design solution can address. Rather, that solution has to take into account an understanding of mental maps, place attachment, generational attitudes, and life experience. Just as importantly, the design should demonstrate sensitivity to the different needs of individuals and communities.

There is no shortage of ideas for a memorial or the reconstruction of Downtown, and city schools, with their eclectic mix of students, are a good place to get a sense of the wide range of views. I decided to begin by interviewing students at a private school in Lower Manhattan not far from the site. I first asked eighth-graders at Grace Church School what they would like to happen at the World Trade Center. Their responses varied as much as those of any group of adults, and were just as imaginative:

A New Yankee Stadium, the World Trade Center towers higher—it shows that we are better than before, even if only three stories higher.

Rebuild taller with a statue as the centerpiece of the place, or maybe there would be a memorial inside.

Have a "Center for the City" to show how we came together—maybe it could be a shelter or a place for counseling people or a YMCA.

Instead of two towers have five buildings in a circle connected by different colored rings and a memorial in the center remembering the people who died—like Ellis Island.

Have a piece of the [original] tower in the memorial. It should be something that shows that we are moving on—not just a cemetery.

I would have a sports field, but something, definitely; as long as they clean it up.

A memorial with all the names, a museum with things about the old building; put up poems and pictures.

Two hollow towers open to tourists with glass elevators and a memorial at the bottom.

I loved the buildings and want them rebuilt. There is an empty space now—it is how people know our city. Rebuild it taller and wider.

A park with a garden and a café where you can put flowers and a place to light candles. And have an event there each year to re-member.

Maybe a regular park, Downtown needs parks. Or a huge memor-ial park where people can walk. A monument of people holding hands to represent the people who helped and the people who died.

These comments suggest a design solution somewhere between memori-alization and forgetfulness, between the desire to build bigger and better and to create a space for coming together, healing, and moving on.

For the eighth-graders, the iconography of the Twin Towers is a crucial component of their mental image of New York City and anchors their ex-istence. Some worry that without the towers, New York is not New York,

or will remain incomplete. Their fragile sense of place attachment requires the physical presence of the buildings to mark the skyline and hold
constant what is right and natural in the world. "Even worse," one girl
says, pausing to take a deep breath, "other people will not know it's New
York if they see the skyline without the towers."

Third- and fourth-graders at Grace Church School painted pictures of
an imagined rebuilt site. Their colorful renderings include many of the
ideas of the eighth-graders, best evidenced in one student's particularly
imaginative drawing of a no-plane zone (see drawing below).

Another student drew tall buildings over memorials that depict the
World Trade Center towers in miniature, and one imagined a Twin Tower
Memorial Cathedral. All places to remember, recover, and reconcile. The
eight and nine-year olds' juxtaposition of buildings and garden, workplace and playground, as well as open gathering space colliding with col-
lapsible modern towers, provides a good sense of the diversity of
imaginations across the city as to what should come next.

So what will happen? Will the city respond by building higher walls
and gates, creating public spaces that are barricaded, guarded, and pa-

Vision for a new World Trade Center, 2001. Picture by Lisa Franklin, fourth grader, Grace
Church School.

trolled? Will the new buildings contain only consumption spaces or have guards that discourage use by minorities and the poor? Or will we build new buildings that allow diverse communities to express their needs and desires in a complex space of reflection and reconciliation? Will our search for security and desire to ward off anxiety continue to create fortified public spaces, rather than a "Union Square" where people can express their grief, love, and humanity?

The diverse range of views expressed by the students compels us to acknowledge that given the rich variety of life here, Union Square is the appropriate ideal. After September 11, everyone went to local parks and plazas. They are the communal centers where people come together and share experiences. The memorials at Union Square demonstrate that the emotional life of New York City still resides on its streets and in its public spaces, and that should be the spirit that infuses any plans for the World Trade Center site.

In keeping with this long-standing, democratic, New York tradition, I imagine a complex space with gardens of reflection and recovery, buildings with memorials and historical documents, as well as places to work and play, and open plazas for people to come together to discuss and disagree in a public environment—a postindustrial plaza where the imagery and imagination of all communities, children and seniors, workers and retirees, residents and visitors, will then find public expression.

Acknowledgments

I would like to thank Carol Collet and the eighth-grade Music and Art class students, and Steve Montgomery and the third- and fourth-grade Art class students at Grace Church School for sharing their ideas and images about the World Trade Center and contributing to this chapter. I would also particularly like to thank Lisa Franklin for allowing us to include her drawing of the imagined World Trade Center site.

A Time for Transportation Strategy

SEPTEMBER 11 MARKED A DAY of great discontinuities in New York and in the United States. When terrorists demolished the World Trade Center, they affected our personal and collective psychology as much as, or perhaps more than, our physical landscape. Five months later, the need to revisit and fix that landscape as well as heal our hearts remains overwhelming. Unlike an ordinary development problem, one where a site is methodically cleared, and architects hold charrettes and competitions to provide for highest and best use, this problem has tragically organic roots and is burdened with a vast symbolic load. Thousands of people from all over the world remain and will always remain there; simultaneously, the site will always be remembered as the place where the golden age of the twentieth century of the United States came to a dramatic closure.

Yet we need a careful strategic response to infrastructure reinvestment and all new investment that occurs from now on. This requires an understanding of the context for such investment, including its most critical elements: the location and densities of the area; the nature of its economic activity; transit accessibility; the agencies, firms, and organizations responsible for the infrastructure; and the ability to finance reconstructed or new facilities. One overarching question is whether the region has the ability to organize and focus in the face of such an immense, multidimensional problem.

For along with the physical destruction of buildings—including over 15 million square feet of office space—was the devastation of the infrastructure supporting movement within and to the WTC area. Two subway stations and one major Port Authority Trans-Hudson (PATH) station were damaged, and two more subway stations were shut down. In addition, tunnels to Lower Manhattan were completely or partially closed, and traffic over bridges to Lower Manhattan was limited. In a region that serves both symbolically and actually as the world's financial capital as well as a major economic center for the region, restoration of infrastructure is the very highest of priorities. With the simultaneous dislocations of office and residential space as well as infrastructure, a carefully thought-out approach is imperative, one that entails the key questions facing New Yorkers:

- Will the site be developed at the previous densities, daytime and nighttime?
- Do economic forecasts project the same levels and types of economic activity and job mixes?
- After we discuss the development issues, what becomes of healing and commemoration? As building commences, will 9/11 fade?
- What will be the role of neighboring areas—New Jersey and the outer boroughs—in the redevelopment?
- What will be the "time factor"? In the time it takes to restore some of the infrastructure (2 to 3 years) or rebuild new office structures, how comfortable will firms and their employees be with their relocations? In fact, how many will have shifted residences, modes of travel, or support activities to adjust to the relocation?
- And, who will ultimately be in charge? Will it be a cohesive, strong, and thoughtful singular agency or group? One that can control development and infrastructure simultaneously? Or will it be an institutional committee? What will be the role of the citizenry? Will they have meaningful input, or will they end up as frustrated taxpayers?

In considering all of this, what we need to remember is that this is not simply about building a subway to have quicker access to Downtown.

This is about a response to an attack on our country that has made us reaffirm our values and think very hard about what our very own city means to us, its citizens, and to the world.

As we begin rebuilding and reinvesting, though, there are a few basic assumptions that provide us good, useful guidance. The first is that the site belongs to the world. After all, the WTC towers were so named, reflecting New York City's place in the global market, and the new development should reflect the world converging on this site as a symbol of freedom, realized aspirations, and global diversity. Moreover, the very dynamics of New York, those that make it unique, have been reflected in its density and agglomerations of activities. Rebuilding the site must account for this drive for agglomeration, business innovation, and great workforce diversity. This intensity and density of economic activity are essential to the definition of New York City.

The site, one of the most valuable in the world, derives much of that value from its accessibility—the ability to move huge numbers of people to the region within acceptable travel times. And the region (and the world) has that workforce. The accessibility comes mainly from rail transit. The Metropolitan Transit Authority (MTA) and PATH served the site, a point of great interconnectivity. Access to the region of the site has stimulated the development of housing, schools, shops, a waterfront park, and other activities. Before September 11, rail transit was facing the limits of capacity, and new capacity, especially from the East Side of Manhattan to Lower Manhattan, was under study. If Lower Manhattan is to remain the focus of financial and related activities and keep the agglomeration effects that characterize it, access must remain high. Reconnecting rail to the regional network becomes a priority of the highest order. But adding in new means of travel, such as new ferry service, must also be done. And while travel time and system reliability were crucial components of rail transit before September 11, since then a new variable—security—must be added to the cost of travel to any new development.

The redevelopment and regrowth process is a dynamic process, and recognizing this is fundamental. In New York City, defining the issues and merging all interests to solve problems related to the issues—especially concerning investment and development—are major ills. The problems don't arise from a lack of innovative thinking. Rather, they emerge from the very diversity of dynamic interests in the city. In a region with a multitude of agencies addressing transportation alone, another multitude dealing with development, and an array of civic groups—official, such as community boards, and ad hoc, such as new "alliances"—where innovation stands is, as they say, where it sits.

It should be clear that a strategic approach to redevelopment is necessary, one that combines very short-term (less than one year), short-term (less than five years), and long-term approaches. The redevelopment of infrastructure is integral to current land uses and activities as well as the evolving land uses that will occur on these three time scales. A strategic process is critical: it examines the availability and characteristics of resources—financial, land, materials, human, and institutional. It also considers the local and regional constraints, identifies opportunities, and allows those in the region to measure the actual progress of redevelopment against a set of regional goals. Such a process can be inclusive, yet bounded, and provide a vision of what can be accomplished to the region.

But who will set the strategy? At this time it is clear that a new agency, put in place just for this task, will have this role. But all in the region want their voices heard, and many sing from different scores. All, however, agree that transportation is critical. Indeed, as will be discussed below, the idea of restoring accessibility is key to the restoration of a certain quality of life to the impacted area, "a New York quality of life." That is, one that stimulates a diversity of activities, promotes opportunities for all, is vital, vibrant, and conducive to an attractive urban environment. On September 10, 2001, Lower Manhattan had something resembling this. On September 12, 2001, it did not. And in January 2002, we are all still trying to figure out how to regain it—although in a new and improved as well as thoughtful and commemorative fashion. That is a tall order! Why is transportation important? It defines the scale! It allows for a particular type of quality of life to occur. A road ending in a cul-de-sac might let a few dozen people an hour travel into a neighborhood. Four or five subway and commuter trains will let 100,000 people an hour come to one place. Just think of the opportunities all this accessibility will provide. That is why restoration of accessibility tops everyone's list of first to-dos.

A Sense of Scale

The loss of the WTC and several surrounding buildings as well as the damage to subway and PATH lines has created significant dislocation. These are dislocations that have had a major impact on firms and businesses, residents, and support services, and on the patterns of behavior that have developed in Lower Manhattan. Nearly 100,000 jobs have been displaced. Over 80,000 transit trips per day must be reconfigured. Four New York City Transit stations were closed because of damage. The PATH railway station, a source of 60,000 daily commuters to the WTC and surrounding buildings, was destroyed. Normal means of intraregional access, by transit, car, and walking, have had to be reconfigured (or forgone). Ferry service, providing trans-Hudson commuting for 15,000 persons before September 11, has doubled since. Trips to the PATH Uptown stations have also doubled, and travel by New Jersey Transit (NJT) to Penn Station has increased by one-third. Many streets have been closed or had their access limited because of the need to allow movement of construction vehicles and maintain public security. And the condition of the WTC site, at this time, is a daily reminder of the suddenness and scope of the tragedy, personalizing it in a manner no ordinary redevelopment project would have.

Through all of this, businesses and firms have had to keep operating. Fortunately, a region as large and dynamic as New York has had the capacity to absorb businesses forced out of 32 million square feet of damaged or destroyed space. Among the larger tenants, 18 percent relocated to New Jersey and 34 percent to Midtown Manhattan, and 46.5 percent remained in Downtown Manhattan. The relocations to Midtown, in particular, have had a staggering impact on NJT and PATH. Both now provide crowded standing room to Midtown. PATH trains to Midtown carry 50 percent more riders at the morning peak than they did prior to September 11. While these are all temporary relocations, the people associated with those firms are learning new travel behavior, and secondary businesses are cropping up to support them. As it will be at least one year before any new or repaired rail infrastructure is in place, and much longer before new structures are operational, questions concerning the makeup of the economic activities in the redeveloped site and workforce that will move into it are critical ones. Simply repairing rail stations so that they restore service to the September 10, 2001, levels assumes activity distributions and levels similar to that time. But discussions and plans concerning new connections and alignments must be linked to focused forecasts of anticipated levels of new economic activity, the labor forces to support it, and their preferred commuting patterns.

Since September 11, 2001, Lower Manhattan has been redefined into three zones. The first is Ground Zero, the 16-acre WTC site. The second is the areas immediately surrounding Ground Zero—significantly impacted by the events, yet viable and operational, such as Battery Park City. The third is the remaining area south of Canal Street. The events of September 11 had great impact on all of the area south of Canal Street because of the effect on infrastructure. In the first several weeks, travel below Canal (indeed, below 14th Street) was severely limited. Later, access to all of the area, with the exception of Ground Zero and its immediately surrounding streets, was restored. However, the perception of difficult access to all of Lower Manhattan, held by residents and tourists alike, has created severe impacts on business and other activities in Lower Manhattan. Meanwhile, even as local services still search for the business to start anew, the WTC site itself has become, of all things, a tourist attraction.

Transportation Accessibility

Rethinking infrastructure investments to support redevelopment of Lower Manhattan underscores both the role and importance of transportation. But as we design a strategic approach to these investments, we need to bear in mind a few essential concepts. The first is that each trip taken by an individual or by a vehicle moving goods can be thought of as an economic transaction. People travel to work, shop, or other activities, and shippers want goods delivered promptly. Transportation represents a cost in those transactions, and infrastructure planners try to minimize those costs (travel time, convenience, and now security) in order to improve the accessibility—and hence the attractiveness—of a particular location. The WTC was characterized by high accessibility; several subway and PATH lines passed through the site. This permitted more than 100,000 persons per day (equivalent to the workforce of Rochester, NY) to arrive there. Conversely, those who had businesses there could be sure that within a reasonable commuting cost, they would have access to a labor force, *no matter what skills were needed*. Ultimately, people came to select housing locations based upon access to the WTC and its surroundings. The ability to bring so many to work in a small land area, using rail rapid transit, built the unique economic engine that worked 24 hours each day, every day of the week. Accessibility, then, is defined by the cost of travel to a location and by the various opportunities (to work, live, shop, etc.) congregated at that location. Obviously, with the removal of both the transportation infrastructure and the opportunities, the accessibility and attractiveness of that location will suffer.

Furthermore, decisions about the land uses, activities, and transportation investments are not separate, but related ones that must be made with the short term in view. But additional questions concern the permanence or stability of relocation. Will firms and individuals make decisions during the period of redevelopment of Lower Manhattan that will permit another relocation to the rebuilt site? Or will these decisions preclude another move? Regional organizations must keep data on these firms as part of the reinvestment process. In any event, the impacts of those decisions will be felt not only at Ground Zero but also in all of Lower Manhattan and ultimately the entire region.

To plan strategically, though, certain key goals need definition, analysis, and resolution. The first is reviving Lower Manhattan's quality of life. Restoring it means restoring safety and security as well as providing outlets for satisfying cultural needs. Just as important is reinvigorating economic growth. New York City went through a period of unprecedented economic growth during the last decade. The economy had slowed before September 11, yet recent growth established Lower Manhattan as the center of the world economy, bringing together finance, real estate, commerce, and support businesses in a dense, stimulating, and highly entrepreneurial environment. Can this level of activity—a level that made the average income of those living and working in this area 50 percent higher than the average in the rest of NYC—be re-created? What will be the roles of high tech and information technology in the restructuring of this local (yet global) economy? What economic stimuli and levels of finance are necessary to hit economic development targets?

The third goal is enhancing connectivity. The WTC site represents a significant portion of the economic activity of the New York region; but it is only a *portion* of the economic activity. There are layers of reconnections that must take place, and this will take place with discussions about fostering other locations—the outer boroughs, New Jersey, Connecticut—for economic growth. The layers begin with Lower Manhattan, reach through complementary and competing locations and end with necessary global connections through the region's airports.

Connectivity is related to one of the central controversies surrounding all of these issues: the debate about how concentrated activity will be in the future. The trend toward dispersal is a natural response to the WTC's destruction. In fact, dispersal is necessary now in order for businesses to sustain themselves. But the underlying rationale for centralization and agglomeration will remain. How close do deal-makers wish to be to each other? In order to sustain a financial industry as inventive and competitive as New York's has been, what clusters must be in place, and what should their magnitude be? There is a current argument that the WTC was not an instant success—so why build to that density? Time, however, has proved this argument wrong. The WTC created an immense amount of floor space, which gradually filled as the economy improved. The measure of its success is the growth of rental dollars per square foot in Lower Manhattan—higher than the suburbs, and higher than across the Hudson. That is because of accessibility and the thriving entrepreneurial culture in Lower Manhattan. The financial deal-makers didn't want virtual meetings and virtual lunches. They wanted, and will want again to touch each other. Remember, in all this discussion, this is New York—not London, not Tokyo, not Hong Kong. And it is open 24 hours a day—because its transportation systems operate 24 hours a day. Only Chicago's transit, in the rest of the world, operates 24 hours a day. Our infrastructure must be as tough as the city itself, which means it will be expensive to build and operate.

None of these goals, it should be said, are isolated from each other;

they are intimately related, which is why infrastructure planning around the WTC is so complex. However, strategic planning can link single objectives and resources into larger, integrated plans, and implement them incrementally over a well-structured period of time. It will take some time to evaluate the economic opportunities of the WTC site and the region, especially during a period of economic downturn. But a period of growth and expansion must be planned for, and levels of activity similar to those prior to September 11 must be factored into the objectives. Good infrastructure clearly stimulates these activities. Since large capital investments take long periods of time, though, the rebuilding of the site must be accompanied by a period of flexible transportation improvements, ranging from the very short term to the long term. The long-term investments must also meet standards of regional economic enhancement.

In the immediate future, local transportation providers might provide a targeted, scheduled, or on-call van, small bus, or other similar service to Lower Manhattan as it undergoes rebuilding and continues to experience job relocations. These might be provided through open-market bidding, perhaps to the Taxi and Limousine Commission (TLC), the city's Department of Transportation, the Mayor's Office of Transportation, or to Lower Manhattan's business improvement district. The service would be dynamic in that routes would change as the construction patterns change. They might also change between daytime (linking Penn Station and Grand Central Terminal to Lower Manhattan or to available subway stops) and nighttime routes. Pedestrian and bicycle routes would also be established, not only to accommodate nonmotorized means of travel but to encourage them, support them, and generally make them appealing. Information kiosks on transportation and how to connect with it (i.e., computer terminals) could be set up throughout Lower Manhattan to provide information on the available systems, making them more attractive. All public transit alternatives, regardless of provider, should take prepaid-fare Metrocards. To ease congestion due to limited street access in Lower Manhattan, new automobile policies could be implemented as trials for long-term strategies to reduce congestion without limiting access. They include limiting single-occupant vehicles on bridges and tunnels as well as new toll pricing and parking policies. Also in the very short term, the movement of goods and trucks must be addressed. Delivery of goods is vital for commerce and daily life, yet no one wants to see trucks. Regulating their access via restrictions on location, parking, vehicle size, and time of day must be a part of the redevelopment strategy.

Short-term projects that are not quite as pressing should address street improvements and the addition of Intelligent Transportation Systems (ITS) to street and transit networks. Bus lanes as well as Bus Rapid Transit (BRT) can begin to address capacity needs as the activity in the region returns to previous levels. Simultaneously, MTA, PATH, and NJT must develop a plan for the reconstruction of damaged lines and the reconfiguration of lines through Lower Manhattan to provide high capacity and to meet movement needs twenty years from now. The MTA can work more aggressively than it does now on modernizing communications and signaling—not only to add capacity and reliability to the system, but also to make the system more responsive in a crisis. The MTA must also review station designs to ensure quick access and egress during emergencies and develop procedures to evacuate stations—especially high-volume stations—during emergencies. Finally, short-term strategies must incorporate reopening closed rail lines quickly to ensure that the expectation of accessibility is not lost.

For the long term, several suggestions have been proposed in the transportation community to improve regional transportation. These include establishing new PATH connections in Manhattan, reconfiguring of the South Ferry and Bowling Green stations, creating a Fulton Street superstation, and building a Second Avenue subway line. Long-term integration of transit lines through easy station transfer points is critical for capacity improvement in the New York City subway system. As well, additional points of access and interchange for NJT, PATH in New York, and at subway stations will provide dramatic increases in regional accessibility. In addition to rail extensions, BRT and light rail additions have been suggested. Finally, there are a number of ideas for addressing street and parking issues. They involve utilizing ITS, new pricing approaches, and strategies limiting single-occupant vehicle access to the most congested areas of Manhattan. Investment in ITS, the applications of computers and communication technologies (EZ Pass is an example), is essential to establish any regional auto and truck use policies.

The Role of Choice

Individuals and firms displaced by the events of September 11 will have a number of choices to make that will eventually impact on the rate of economic regrowth in Lower Manhattan. But activity decisions are also transportation decisions, and levels of transportation investment must coincide with the rate of, location of, and scale and density of downtown economic regrowth. Individuals can relocate from jobs or housing, or both—or neither. Firms can relocate temporarily or permanently, or restructure with only portions of the firm allowed that choice.

Transportation plays a critical role in all of these choices. Most people use some aspects of cost (including the price, travel time, and reliability of the trip) and often weigh this against their income. There is a threshold at which the cost is too great and the trip is forgone. To many after September 11, new commuting patterns have come at increased cost (in terms of time, transfers, and comfort).[1] In a temporary situation, this cost can be absorbed. But if it's permanent, the worker might want to either find a new job or relocate to reduce the cost of travel. There is another variable to factor in now, however: security. Are the subways safe? Security from terrorism is a new variable in mode choice, and will be considered in job locations. Finally, of course, firms will locate or relocate where they believe they have access to the labor force that provides the skills they need. A 45-minute circle centered on Midtown Manhattan (Penn Station) or the WTC would enclose millions of workers. A similar circle around Jersey City or a suburban location would encompass far fewer workers. This again is an illustration of the importance of accessibility.

As redevelopment plans emerge from both community and public

---

[1] Anecdotal evidence from many who worked at WTC for the Port Authority indicated that 45-minute commutes from Upper Manhattan, Bronx, Queens, and Brooklyn to the very accessible WTC site had become arduous one-and-a-half-hour commutes to Jersey City.

agencies, a fundamental question will have to be addressed. The subway stations that were damaged and taken out of service were part of a system designed over a hundred years ago. For the needs of 2020 and beyond, how should system reinvestments be made? Should all stations be opened as before, or should new connections and links be made to respond to emerging or desired travel patterns of the future? While there is a limited number of choices that can be made, each investment strategy is expensive: some will require joint agency participation (MTA NYCT, PATH, NJT), and some will require integration into new types of development. All will have to be consistent with infrastructure development and system modernization occurring elsewhere in New York (e.g., the East Side Connector, the Second Avenue Subway line, and the sustained growth of NJT).

Current Plans and Strategies

There have been several responses and plans to reinvest or redevelop infrastructure post–September 11. These include plans put forth by community and professional coalitions as well as activities undertaken by government organizations and the private sector. Generally, the immediate concerns were to provide transportation to those who still had jobs located in the region and to those whose jobs had been displaced but relocated within the region. Estimates of required time to restore the greatly damaged Cortlandt Street and WTC PATH station to full use are from one to three years. However, the MTA has devised a strategy to reopen two stations south of the WTC (Rector and South Ferry) in order to restore much of the transit service available before September 11. The Port Authority has already commenced planning to reopen a PATH station (perhaps temporary) at the WTC site within two years. It has noted that such a move will not preclude any of the more ambitious plans for PATH in Manhattan. In fact, to meet the needs of what some regional bodies call for—a superstation connected to Fulton Street—the PA will connect its rebuilt WTC station to Fulton with moving walkways.

It is interesting that it has taken decades to begin utilization of underground space in a serious way. Visitors to Tokyo, Toronto, and Montreal all see how underground spaces, in addition to being connections, can be activity centers themselves. The spaces in Montreal and Toronto, starting as cold weather walkways, have become year round economic venues. In Tokyo, land on, above, and below ground has value. Ideally, it will be possible to have from 100,000 to 200,000 people per day go through the proposed connection in Lower Manhattan. A great location for a newsstand! And such a site can be a sample of what can be done through planning elsewhere in NYC.

For commuters from New Jersey (the PATH users), ferry service, primarily supplied by New York Waterways, has significantly increased, now carrying more that 40,000 persons per day to Lower and Midtown Manhattan. PATH service continues from New Jersey to Midtown, where commuters can transfer to the reconfigured subway system. The reconfigured service now reaches most of Lower Manhattan outside Ground Zero and areas immediately south to the tip of Manhattan. Bus service has been redesigned to serve residents of Battery Park City, connecting them with active subway stops. But Battery Park City, adjacent to the WTC site, remains somewhat isolated and relatively inaccessible.

To address the need to sustain access to the WTC site for emergency and construction vehicles as well as the potential growth of congestion on an abbreviated street network, policies were put in place to restrict automobiles from the area or to limit auto use in the area. These policies included closing adjacent tunnels for an extended period and, more recently, limiting single-occupant vehicle use through certain tunnels and bridges at specific times of the day. The success of these policies may lead to their permanent adoption in broader transportation contexts in the future.

Long-term strategies have been proposed by a number of regional coalitions, community groups, and professional groups. The most ambitious would have new subway development beyond the reopening of the damaged stations on NYC Transit and PATH. Some would extend PATH's reach in New York City beyond the WTC site to the Wall Street area (Fulton Street), where it can connect with more subway routes. Additional plans would have a South Ferry Station reconfigured and extended to meet with the Bowling Green Station, again providing improved transfer facilities. But all long-term strategies are ultimately linked to the new land uses and activities, as well as to the regional will and ability to find financial resources to support them.

Institutions and Financing

Finally, these strategies must be put into the context of the organizations that develop, implement, and operate them. It may be that the structure of the existing transportation agencies may have to change, or new agencies be put in place to deal with the new complexities of twenty-first-century planning and delivery of infrastructure. There are signs of interagency cooperation that has already occurred to ensure that accessibility improvements are made. Much of the current cooperation—for example, between the city Department of Transportation, the MTA, NYC Transit, and the Port Authority—revolves around security issues. But the ability to use Metrocards on ferries and the discussion between MTA and NJ Transit concerning single-fare travel are major steps towards improving accessibility through system integration.

A new superagency, the Lower Manhattan Development Corporation (LMDC), created by New York State to be a subsidiary of the Empire State Development Corporation, is an example of inventing a new institution to operate without the limitations of special-purpose state and city departments and agencies. It will have broad powers over a great part of Lower Manhattan, including eminent domain and the issuance of debt. At the time of writing, the LMDC is committed to restoring a high level of economic development and activity at the WTC site.

And, of course, there is the pressing issue of financing regional redevelopment. After the federal money is gone and bond limits met, there will still be monumental needs. Innovative financing will be needed, coupled with innovations in building and operation infrastructure. Examples in Europe and Asia exist already. These two aspects, institutional change and finance, will have a critical impact on the rebuilding of the WTC site and on regional development and growth. For the time being, the most common sources of funding will be subsidies and grants. The federal government, through the Federal Emergency Management Agency (FEMA), has promised funding to repair or replace existing infrastructure facilities.

But beyond those, local funds will be needed. While some of this can come from the issuance of debt, there is a limit to the capacity of the region to issue bonds. There will be innovation in this realm, especially in the form of public-private agreements. These might center on real estate transactions. Developers might be given certain rights to invest in or build stations integrated with commercial, retail, and residential centers. Revenue from road and bridge tolls might be used to cross-subsidize transit development. The need for substantial capital investment over the next five-to-ten-year period will put pressures on the region to investigate and adopt new financing techniques and to stimulate change in the investing agencies so that they utilize such methods.

What Will It Take?

All of this seemingly self-evident strategic thinking will not, in itself, start the reinvestment process. The process will have two lives. The first is that pumped into it by the Lower Manhattan Development Corporation. It has the clout and the political go-ahead. It has the ability to make decisions on the basis of board votes, using members' best judgment or the input of all who might want to participate with ideas. It has to balance the pressures of satisfying very local demands for restoring quality of life and pure economic pressures for large-scale, rapid growth. Ironically, the infrastructure will support either. But the redevelopment authority is by definition focused on the WTC site and surroundings.

The second part of this process is creating an environment to invest in infrastructure renewal and rebuilding in a much more progressive and twenty-first-century manner. An alphabet soup of infrastructure agencies, ranging from local to federal—NYSDOT, NYCDOT, MTA, NJT, NYC-DOP, USDOT, EPA (controls transportation through air quality), NJ-DOT, PANY&NJ—and private operators have a chance for a unique sort of cooperation. But it is incumbent upon them to settle on the best ways to facilitate every form of transportation—from door-to-door trips to truck shipments of goods. Furthermore, how can we encourage and protect walking when, in fact, all of these agencies deal primarily with motorized transport? In an area as small but dense as Lower Manhattan, most journeys can be made by foot! The agencies, all historic products of the transportation planning revolution of the twentieth century, have the first real and immediate infrastructure challenge of the twenty-first century. Investments made today will have impact for the next fifty years and will only begin to hit their stride in 2020. That is the lesson we have learned from all of the big twentieth-century investments. First infrastructure, then development or redevelopment. A textbook picture of how that occurs is available by simply looking at New Jersey across the Hudson. And if you look in twenty years from now, it will look like the West Side of Manhattan. Lower Manhattan can be a case study of new institutions arising out of old. Centralized control of motor vehicle movements and transit; major transfer points among modes; simple and universal fare payment devices; and information on infrastructure performance, available in real time, everywhere. This includes on the Web, on your PDA or mobile phone, and at kiosks everywhere (where you can also purchase or charge transit fares or auto parking and use fees, or call taxis).

We have a unique opportunity. New Yorkers are used to thinking big—except when they want their institutions to do something. We have the brains, energy, ideas, and the right technology to rebuild the Lower Manhattan infrastructure as a working model for the world of how to do it right in the global city. But if we rebuild the last century's systems, we will greatly shortchange ourselves. It is not money that stops us; we are the richest country in the world. It is not institutional inertia that stops us; we can invent, and have, here, new institutions. It is not NIMBYism that stops us; neighbors can be bought off. If by chance we do fail, it will result primarily from an unwarranted defeatism—that is, our deeply rooted belief that the people who built the George Washington Bridge as well as all the tunnels and subways have all moved on to the suburbs and abandoned the city. But the city will continue to change in a positive direction if we remain highly aware of all our options and manage them well. As we rebuild and reinvest, then, the picture in mind should be of a New York of September 10, 2021, and not the New York City of September 10, 2001.

Enduring Innocence

THE AL QAEDA NETWORK has been likened to the Barbary pirates, the North African corsairs of the early nineteenth century. Like the Al Qaeda, the pirates received temporary
support from various regimes and were brought to justice by a
coalition of nations believing themselves to be ethically and
technologically superior. Yet our own politicians and businessmen sail a strikingly similar pirate sea, slipping between legal
jurisdictions, leveraging advantages in the differential values
of labor and currency, brandishing national identity one moment and laundering it the next, using lies and disguises to
neutralize cultural or political differences. Indeed, the sea has
become a favorite metaphor for globalization, not only for its
transnational networks and shady offshore activities but for
the potential political power of its citizens to identify responsibilities as well as opportunities in global currents. On September 11, a terrorist network inhabiting this familiar,
denationalized territory declared architecture—as building
structure and as urbanism—to be a primary adversary or rival
on the pirate seas. By translating the static envelopes of buildings and the conventions of urbanism into an apparatus of
war, the attacks revealed a latent political agency for architecture. Buildings and cities, now clearly more than just national
icons, were suddenly cast as active players in global markets
and political conflicts, to be assessed for their weakness and
resilience, their potential for aggression, and their ability to
collude and resist.

Warfare and piracy oscillate between symmetrical, complementary, and reciprocal organizations. These expressions, borrowed from anthropology, attribute agency to organization and apply broadly. They facilitate a correspondence between architecture and warfare and help situate the discipline within a broader political context. In general, a symmetrical organization is competitive: rivals clash with and mimic each other over similar goals. A complementary organization is hierarchical: one side is clearly more powerful, and the other submits. Both symmetrical and complementary organizations can generate a fragile stability, but they can also produce rivalry, insurgency, and violence. A reciprocal organization, on the other hand, neutralizes a competitive symmetry or a complementary hierarchy by surrendering interest in the fight, multiplying both enemies and friends, and entering into a more complex web of cooperative engagements. Moreover, in our common cultural assumptions about organizational complexity, we understand that an abundance of information, redundancy, and error—whether it is in networks, genetics, or architecture—creates the most robust organizations, while the exclusion of information weakens an organization's resilience. Similarly, all three organizational types permit or exclude information to varying degrees. Symmetrical and complementary organizations must exclude some information to maintain their order, while reciprocal relationships are not only the most stable but the most information-rich.[1]

Productive piracy tends towards the reciprocal, countering symmetrical entrenchment with cooperative structures. The evasive corsair moves from advantage to advantage, often in disguise, aggressing to keep the wheels of exchange spinning but stopping short of a war that would arrest the vital flow of exchange. Open warfare signals piracy's failure: a direct, symmetrical confrontation with the enemy, it destroys the pirate's rich relationships with adversaries and coconspirators. In this weakened, less complex organization, an escalating rivalry and mimicry drive adversaries either to destroy each other or to self-destruct. A sense of righteous inno-

---

[1] Anthropologist Gregory Bateson developed this theory of organizations. His eccentric project searched for shared organizational principles in networks, mechanisms, animal behavior, or global politics as a means to extend the territory of his own discipline.

cence—itself a product of restrictions on information—often accompanies this shift to violence and warfare. The same expressions of political agency apply to architecture and urbanism as well, when these are understood not merely as building envelope and master plan, but as organizations, capable of a kind of piracy responsive to and responsible for a broader political context.

Bush and bin Laden, the two sons of mid-century oil privilege, engage in a mimetic symmetrical conflict within which they cast buildings and cities as major players. The righteously indignant Bush administration has repeated over and over that our adversaries live in caves. The Al Qaeda network, the primitive nomad, will be brought to justice. Caves are Stone Age and the World Trade Center is a gleaming specimen of technological superiority. Bin Laden also positions himself on the side of absolute righteousness, forced by increasingly intolerable circumstances to give battle to the other. To Al Qaeda, though, caves are complex organizations and the World Trade Center is a vulnerable primitive. But both the United States and Al Qaeda must fight on the other's turf. The United States must wage war on an exotic battlefield against the shockingly repressive regime harboring Al Qaeda. Al Qaeda fights in the shadows of everyday life in America, from suburban hotels in greater Tampa to flight schools outside Minneapolis to, finally, the brilliant morning skies over Lower Manhattan. Yet from this symmetrical formation—and consistent with its denial of information—emerge the most resolute cries of innocence. Claiming that we did not ask for this mission, the Bush administration sends "heroes" to fight the evildoers. Bin Laden's early comic denials and his theatrical renunciations of Western culture exhibit the same degree of righteousness.

The United States constructs its sovereignty by constantly switching between symmetrical and reciprocal stances. The symmetrical pursuit of war in Central and South Asia, while currently a primary concern, is merely ancillary to the more complex reciprocal organization of the global coalition that the United States administers. The coalition conducts additional campaigns on the information front, with both allies and potential enemies and on behalf of other, related agendas—protecting commercial relationships and oil interests in the Arab world and the

Caspian basin. Al Qaeda also switches between the symmetrical and reciprocal. Its operatives enter into temporary agreements and adopt temporary disguises to operate in the thick of America's most banal cultural offerings while also preparing an assault on that culture. Orchestrating young men's lives for both assimilation and warfare, Al Qaeda decides when to schedule their enrollment in school and when to schedule their suicide.

This oscillation between the reciprocal and symmetrical, between cooperation and direct warfare, marks the most dangerous waters in the sea of globalization. Attending declarations of righteousness and innocence amplify the potential for violence by withholding information. By seemingly maintaining a healthy, open network of associations yet operating via a secret agenda, a reciprocal organization serves as an accessory to strengthen a symmetrical order and its periodic episodes of destructive violence.

In these treacherous oscillations between reciprocal and symmetrical architectures, between the expansions that admit new information and the contractions of righteous war, the World Trade Center both provoked and succumbed to warfare. The goal of terrorists is to prompt their enemy's own self-destruction, and in this symmetrical relationship, the World Trade Center was both combatant and intended instrument of suicide. The attacks tested each building's power to resist, cooperate, or surrender, and from not only a structural but an organizational perspective. More important, the attacks revealed a broader transnational network of geopolitical influences that are typically considered to be beyond the reach of localized urban planning.

As a building, the World Trade Center was a spectacularly vulnerable adversary when symmetrically aligned with its attackers. Yamasaki's design for the World Trade Center was borne into a complex urbanism of weak building codes, real estate schemes, and urban master plans, a weblike, reciprocal organization, filled with productive lies, scams, and disguises. Contemporary journalism often focused on the building's "smart" elevator network as just one of its intelligent building systems. But while the building appeared to be a giant, nondescript computing unit, filled with an efflorescent network, unlike the elevator technologies within or the global urbanity of the city outside its doors, the simple volumes contained in its spatial envelope had a much less intelligent organizational repertoire. To maximize rentable area, the floors were almost completely segregated, connected only by thin strands of exit stairways. Finally, the buildings rose from their urban context via a symmetrical, competitive urge: to offer the largest volume of commercial real estate in the world's tallest towers.

As an organization—or a network—of spaces, the World Trade Center was a serial rather than a parallel arrangement. Like serial computing systems, in evacuation, the World Trade Center required slow, sequential routes of circulation rather than the simultaneous and reciprocal points of contact and communication associated with parallel computing. The Pentagon, of near-equal capacity and only surpassed in size by the World Trade Center at the time of its completion, was a much more resilient organization because it contained multiple redundant pathways of entry and exit close to the ground. Evacuating the building took minutes rather than hours. The World Trade Center also approached the limits of spatial efficiency that the supertower model imposed. A thickening core designed to deliver quickly a population of people to the upper floors reduces the rentable floor area that funds the entire enterprise. (Builders of more recent supertowers, working within similar limits, would make the elevator core the means of fire-safety egress and dedicate a separate bank of elevators for firefighting, thus extending the intelligent redundancies of the infrastructure to safety issues and facilitating better communication between floors.)

The World Trade Center collapsed in explosion and implosion, detonation and self-destruction. Yamasaki's own Pruitt-Igoe housing towers in St. Louis, demolished in 1973, just a year before the completion of the World Trade Center, marked the beginning of a number of intentional implosions of high-rise housing towers all across the country, housing that had almost self-destructed in the face of adversaries like drugs and crime. Any user, dealer, criminal, or maintenance problem affected the entire tower through the core, the unavoidable space of circulation. The towers were exactly the kinds of structures that an epidemiologist would regard as highly susceptible to contagions. Like these towers, the World

Trade Center organization was so singular, so impossible to partition, that any negative influence could possibly unleash a deadly epidemic. The reductive and symmetrical organization, lacking resilience, with no means to dissipate disturbance, only enhanced the potential for catastrophe.

The collapse also identified a constantly expanding network of remote sites implicated in building and urbanism as well as global politics, sites that avoid direct, symmetrical conflict but have the potential to leverage rich reciprocal relationships. For example, in both economic and military theaters around the world, oil, the Stone-Age currency, still dominates. The wreckage of the towers continues to outgas toxins, releasing information about our choices of materials and about our many industries tooled for the use of petroleum products. The acrid air perhaps even serves as an indirect reminder of the far-flung oil economy and the attending finance capital that was concentrated in the towers. In a sense, the oil fight is one of America's symmetrical squabbles over a limited resource, and we continue to pay dearly for it. Yet the shortsightedness of America's energy policy marks an expansive territory for change, as yet only postponed. This is territory for architectural invention with new materials and technologies that will, by necessity, eventually engage a broad reciprocal substrate of the global economy.

While architects busy themselves with new urban plans for rebuilding Lower Manhattan, the most profound effect on global urbanism and global politics will reside in the ingenious adjustments we make—or don't make—to the habits, practices, and desires that structure our petroleum-fueled culture. The righteous fight and the pallid aesthetics of some sustainability efforts that directly address the evils of energy wastefulness with a symmetrical construction exclude opportunities for leveraging a more complex outcome within the productive piracy that is urbanism. Similarly, when some caution against any collusion with global capital, against "riding the waves" on the vast seas of globalization, they may be overlooking a more circuitous, lateral path toward social transformation, one that relies on indirect persuasion, collusion, and conspiracy as well as invention.

Architecture inserts the most immovable objects in the flows of world trade, but it also commands and modulates vast networks of materials and labor in these flows. Architects rarely consider the political power of these larger networks and often feel they are exempt from any complicity in them—even in the largest-scale, most capital-intensive projects. We still build primarily according to instructions from a hierarchy of clients who, to protect themselves, oscillate between symmetrical and reciprocal responses to culture. The architect is only following orders, we say. Architecture does not believe this to be a dangerous stance, but one that produces innocence.

In collapse, in surrender, overwhelmed by a sea more abundant with information than before, the World Trade Center leaves behind the possibility of a reciprocal architecture. The avalanche, as a kind of suicide, was not agonizing or torturous, but terrifying, vicious, and instant. The building's self-destruction was the cruelest way to release information, in crisis and helplessness. The incineration, the rapid transformation to ash, and the simple release of surrender produces an ineffable and frightening comfort. That arresting condition might trigger the elegiac, a righteous, symmetrical defense of our tallest building, and claims of exemption and innocence that ignore the real power of the networks made so visible on September 11. Or the attacks might manifest a global practice of architecture with political stakes, a cagey effectiveness that replaces innocence with ingenuity and piracy.

The Center Cannot Hold

IN HIS FAREWELL TO OFFICE, Rudy Giuliani, standing in St. Paul's Chapel, adjacent to the World Trade Center site, declared, "I really believe we shouldn't think about this site out there, right behind us, right here, as a site for economic development. We should think about a soaring, monumental, beautiful memorial that just draws millions of people here who just want to see it. We have to be able to create something here that enshrines this forever and that allows people to build on it and grow from it. And it's not going to happen if we just think about it in a very narrow way."

Giuliani's speech reminded me of Eisenhower's leave-taking from the presidency, in which he warned the nation against the growing antidemocratic power of the "military-industrial complex." In both cases, the cautionary appeals resonated because of their sources, one from a military man and architect of the Cold War, the other from a mayor whose leadership favored planning by the "market."

Giuliani's heartfelt call for restraint ran counter to the business-as-usual approach that has dominated official thinking since the tragedy. This has included obscene ambulance-chasing on the part of the architectural community and robust talk about responding to the terror by rapid rebuilding, bigger than ever. The Lower Manhattan Development Corporation, empowered to decide the future of the site, is headed up by a patriarchal ex-director of Goldman Sachs whose credibility seems untainted by the spectacle of his own firm migrating over the river to Jersey. With the exception of a single community representative, the board is comprised of the usual business crowd. Their initial consensus seems to favor the construction of a very large amount of office space on the cleared site of the fallen towers, with the memorial simply a modest component. And rumors grind on about the working drawings already in CAD at SOM.

Fortunately, the competition for authority over the site is both structural and complex. The Port Authority, Larry Silverstein (the ninety-nine-year leaseholder), the Development Corporation, the federal, state, and city governments, survivor groups, the local community, the business improvement district, the Battery Park City Authority, the Metropolitan Transit Authority, and myriad other civic and private interests are jostling to be heard and influential. If nothing else, this fog buys time for contention and for the serious consideration of alternatives.

What is clear is that despite the currently soft market, some of the 15 million square feet of lost space needs to replaced sooner rather than later and Downtown's dysfunctions repaired to allow the city's economy to reestablish jobs and networks lost in the attack. And the eventual need is not simply for replacement space: the "Group of 35"—a business-heavy organization chaired by Charles Schumer and Robert Rubin—has predicted (in a report about the future of the city's commercial space) that an additional 60 million square feet of office space will be required by 2020.

The question is where to put it. Some will clearly go to Lower Manhattan. The relationships of propinquity that form the social and spatial substrate of commerce and culture downtown must be quickly reestablished by the restoration of infrastructure and the addition of new space. Railroading the restoration of the status quo by looking at the site as no more than its footprint, however, guarantees that we learn nothing from the tragedy and let the opportunity for better thinking slip away.

My studio is downtown, not far from the Western Union building at 60 Hudson Street, known to architects as the home of the New York City Building Department. Since September 11, this building has been the subject of unusual security, surrounded by concrete barriers and half a dozen police cars. It appears to be the only site outside the confines of Ground Zero to enjoy this level of fresh protection, and the reason seems to be the building's long-standing role as the nexus both for telecommunications cables coming into New York City and for trunk lines to the nation and the world, a logical next target for terror, according to some scenarios.

Ironically, a system at the core of urban disaggregation depends on the joining of huge dispersed networks on a single site. The currently dominant pattern of our urbanism—enabled by the kinds of instantaneous artificial proximity that the electronic network of phones, faxes, e-mails, and other global systems allow—is the rapid growth of the "edge city," a sprawling realm that has become the antithesis of a traditional sense of place, the location to which security-conscious firms are now increasingly impelled to retreat.

Of course, this "nonplace urban realm" is the result of more than new communications technology. The suburbs were fertilized by massive government intervention in highway construction, by radical tax policy, by changes in the national culture of desire, by racism, by cheap, unencumbered land, and by an earlier fear of terror. The prospect of nuclear annihilation that made urban concentrations particularly vulnerable was on the minds of many planners during the Cold War, both in the United States and abroad. The massive deurbanization in Maoist China, for example, was the direct result of nuclear anxiety. The dispersal facilitated by the interstates—our erstwhile "National Defense Highway"—was likewise more than simply good for General Motors: both the company and the United States were playing at the same stratego-urban games.

Whatever the causes, however, the effects of this pattern of urbanization were in many ways antithetical to the presumptions behind the space of Lower Manhattan and its kith. Here, concentration has long been considered crucially advantageous. The possibility of conducting economic affairs face-to-face, the collective housing of related bureaucracies and businesses (the famous Finance Insurance Real Estate [FIRE] sectors that make up the majority of business downtown), the dense life of the streets, the convenience of having everything at hand, are the foundation for the viability of the main financial district for the planet. Its characteristic form—the superimposition of skyscrapers on the medieval street pattern left by the Dutch—has given Downtown its indelible shape.

A Plan For Lower Manhattan

Ground Zero

Ground Zero is a sacred place, no more suitable for building than Gettysburg or Babi Yar. This proposal suggests that a memorial be begun as a berm surrounding the site, providing protection for site work and a platform for public viewing. Eventually, this berm would become permanent and within it a shallow green crater—filled with earth from every country—would be an Elysian Field in perpetual memory of the fallen. At the very least such a memorial might serve as a place-holder for an international competition to determine the future of the site.

West Street

West Street is an intimidating barrier, dividing Battery Park City from the life of the island, forcing pedestrians onto bridges. Rebuilding West Street underground from Harrison Street south would take through traffic off local streets, permitting a green seam to stitch the fabric of Lower Manhattan back together.

Transport

The proposal replaces a portion of the yacht harbor at the World Financial Center with a covered ferry terminal linked to a rebuilt Winter Garden. This, in turn, would be connected to the PATH and subways to create a continuous intermodal transit hub and shopping center beneath the site. This space might take the form of a hypostyle hall with 3,000 columns—one for each victim.

Greenfill

Lower Manhattan is an ideal place for walkers, and this plan proposes a substantial reclamation of street space for pedestrian and "slo-mo" circulation. Ground Zero becomes a point of dissemination for this new network, which is meant as a model for the recovery of the public realm from the car. In particular, the plan proposes initial appropriations along Broadway, Greenwich Street, Chambers Street, Fulton Street, John Street, Liberty Street, and Vesey Street.

Hydrologic Fill

The dead zone created by the squared-off configuration of the north end of the Battery Park City landfill is redressed by additional fill which smoothes, aerates, and cleanses the flow of the Hudson. The fill also provides additional recreational space in an area of New York poorly provisioned with it.

New Buildings

Although this scheme is predicated on disaggregation, the construction of replacement office space on sites scattered around the city, there are nonetheless a number of building opportunities throughout Lower Manhattan, a number of which are indicated. In addition to sites on *terra firma*, the East Side of the island also provides many opportunities for pier-based construction and the revival of an active commercial waterfront. This plan suggests office space, housing, and marine activities either as low-rise or as skinny towers on new and reconstructed piers along the southern rim of the island.

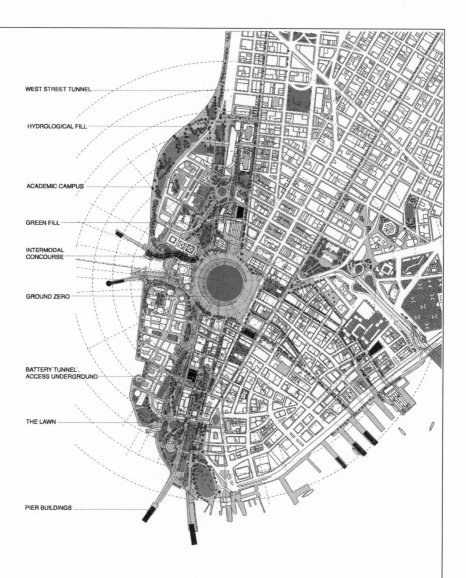

WEST STREET TUNNEL

HYDROLOGICAL FILL

ACADEMIC CAMPUS

GREEN FILL

INTERMODAL
CONCOURSE

GROUND ZERO

BATTERY TUNNEL
ACCESS UNDERGROUND

THE LAWN

PIER BUILDINGS

Downtown Campus

There is a concentration of academic institutions, including Borough of Manhattan
Community College, St. John's University, Stuyvesant High School, and several interme-
diate and elementary schools that form the core of a potential downtown campus. The
suppression of West Street offers the opportunity to create a series of synergistic
quads, linking this collection of schools into a continuous place and providing the community with additional green and athletic space.

The Lawn

Burying the Battery Tunnel approach and demolishing the large municipal parking
structure creates an opportunity for a dramatic new civic space, extending from Rector
Street to Battery Park and offering sites for new development along its edges.

Any changes reconstruction brings must deepen this formal singularity, expand the possibilities of exchange, and broaden the mix of uses supported. While it is now a bromide to apologize before suggesting that the tragedy can be turned to advantage, to have positive consequences, we must find a way to make things better. Life downtown is going to be refounded, and the imperative is to do it well. Perhaps we can reply to the terror by showing that evil acts can have unintended positive consequences.

In fact, the enormous disruption in the life of the city has already had a number of constructive effects. Here in Tribeca, not far from Ground Zero, traffic is dramatically reduced on local streets, the polluted sewer of Canal Street is suddenly tractable, and deep civility abides many months later. The emergency carpooling and limited access instituted as the result of the disaster are equally positive and important contributions to a sustainable urban ecology.

The radical act of the terrorists opens a space for us to think radically as well, to examine alternatives for the future of all of New York City. It is no coincidence that we have constructed a skyline in the image of a bar graph. This is not simply an abstraction but a multiplication, an utterly simple means of multiplying wealth: where land is scarce, make more. Lots more. There is a fantasy of Manhattan as driven simply by a pure and perpetual increase in density. But while our dynamism is surely a product of critical mass, not all arguments for concentration are the same. Viewed from the perspective of the city as a whole, the hyperconcentration of the World Trade Center was not necessarily optimal by any standard other than profit, and even that proved elusive.

Density has a downside in overcrowding and strained services, but this is not necessarily the result of the hyperscale of any particular building. More critical than specific effects on the ground are the consequences for densities elsewhere. While the anxiety over corporate and population flight to the suburbs comes from a general fear of both economic and social losses, the all-eggs-in-one-basket approach slights other areas of the city themselves in need of jobs, construction, and greater concentration. Manhattan's gain has been the other boroughs' loss: the rise of the island's office towers historically marks the decline of industrial employment throughout the city and has obliged the respiratory pattern of one-directional commuting that marks its life. A new form of producing wealth with new spatial requirements has over the century completely supplanted its predecessor.

With thousands of jobs already relocated out of the city, a solution to the "practical" problems of reconstruction can and must engage possibilities well beyond the confines of the downtown site. While the billions that will be available for new building—from insurance, from federal aid, from city coffers, from developers—are certainly needed to restore health to the enterprises formerly in or servicing the World Trade Center, it seems reasonable to question—given the probable level of this investment—whether such massive expenditure should be focused exclusively here rather than throughout the city at additional sites of need and opportunity, places development could transform.

The majority of New York City's population and geography does not lie in Manhattan: the island comprises only 8 percent of the city's land area and 19 percent of its inhabitants. Moreover, according to the 2000 census, the residential growth of the island since 1990—slightly over 3 percent—lags far behind the explosive enlargement of Staten Island (17 percent) and Queens (nearly 15 percent), and the dramatic increases in the Bronx (10.7 percent) and Brooklyn (7.2 percent). Manhattan, however, remains the city's economic engine, producing up to 67 percent of its jobs and 46 percent of its retail sales.

The effect of these imbalances on the ground has fundamentally reshaped the city. The great infusions of capital and the artificial fortunes of the last decade have propelled the price of real estate in much of Manhattan to the stratosphere, accelerating the flight of the middle class and the poor and making Manhattan increasingly monochrome. We continue to revere our island as a place of thick, urbane interaction and cling to the fantasy of the great mixing engine of difference, of a place with many quarters housing many kinds of people. Increasingly, however, the differences in Manhattan's neighborhoods are merely physical. This uneven development and accelerated metamorphosis have had dramatic effects, distorting the character of our urbanity decisively.

Here in Tribeca, we are at the end of a familiar cycle in which a neighborhood moves from a mix of warehouses, manufacturing, offices, and housing, to an "artistic" neighborhood, and now to the climax form of gentrification, an extreme-high-end residential *quartier*. The corollary is that the jobs and people formerly employed here have either been eliminated or moved elsewhere: to the Hunts Point Market in the Bronx, to low-wage environments offshore, to the suburbs, or to the new bohemias of Williamsburg or Long Island City. We have scrupulously preserved the architectural character of Tribeca but at the expense of its human character.

With the exception of Chinatown, Manhattan south of 110th Street has become a faded mosaic of former ethnic enclaves and cultural variety. Increasingly, the city's ethnic and cultural quarters are being solidified outside the borough, in Flushing, Greenpoint, Dumbo (Down Underneath Manhattan Bridge), or Atlantic Avenue. Although the city remains a beacon for immigrants—both from abroad and at home—the sites of intake and expression are not what they were, preserved to death. Manhattan is ceasing to be a place to get a start and becoming inhospitable to striving, less and less like New York.

But big changes can also suggest big opportunities for burgeoning neighborhoods struggling to find form or merely to keep up. Not all disaggregation leads to sprawl. Better, perhaps, to call it reaggregation, but it is also a notion that can be useful in cultivating character and encouraging development within more traditional, compact cities like New York, itself the central place for an enormous region. The point is not to make New York more like Phoenix or Los Angeles, but to make the city as a whole more like New York.

Because of its dynamic population and superb movement infrastructure, New York City can become a model of a new kind of polycentric metropolis, with Manhattan remaining its *centro di tutti centri*, its concentrated vitality unsapped. In fact, Manhattan is itself already polycentric: the disaggregation represented by, for example, the easy movement of financial and legal services firms from Downtown to Midtown in recent years suggests that there is a certain fluidity to the idea of proximity within the city, that convenient movement and strong local character can substitute for immediate adjacency within an overall context of density.

Reinforcing New York's special polycentricity would return the city to something of its pre-twentieth-century character by restoring a network of autonomous, comprehensible places. Such a "village" structure—the origin of the great city of variegated neighborhoods—is again made possible by the technology behind the ephemeral and flexible nets and flows of the twenty-first century. Because it is aspatial, this malleability need not simply lead to generic sprawl but can fit within—and reinforce—pre-existing infrastructures of neighborhood difference.

Cultivating this "natural" polycentricity would multiply opportunities for more self-sufficient neighborhoods where people walk to work, to school, to recreation, and to culture. Such places would also satisfy many of the needs that impel people to seek the densities and economies of the suburbs and edge cities. By regenerating local character, the energy of intracity reaggregation could reinforce the expressive singularity of each of these neighborhoods to which its energies were applied—the Asian flavor of Flushing, the Latin-American atmosphere of the Bronx Hub, the African cultures of Harlem.

This would be an advance on the wing-and-a-prayer style of current planning, in which good intentions are simultaneously frustrated by imprecise plans and the absence of economic drivers to set them in motion before changing times render them irrelevant. By joining physical planning to direct investment and to zoning and economic incentives, we can redistribute uses to a set of centers outside Manhattan where land and transit connections are available and economical, places like Flushing, Jamaica, Queens Plaza, Sunnyside, the Bronx Hub, St. George, and downtown Brooklyn, among others. These sites—also identified in the report of the Group of 35—are not mysterious either in their needs or their suitability.

Planning comprehensively could help assure the mixed-use character of these places by including residential construction matched to the numbers of new workplaces, a pattern that has already begun downtown, where substantial office space has actually been eliminated by conversion to residential use. Indeed, in the last ten years, forty office buildings have been converted to residential use downtown, part of an 18 percent population growth in the area below Canal Street. The sense of locality that grows

from a well-finessed mix would be further reinforced by the decentralization of cultural growth (the City Opera, the Guggenheim, the Whitney,
the Met, and the Jets are all seeking space) and by encouraging the development of new cultural, health care, educational, and commercial institutions to enhance the variety and life of these neighborhood centers.

It is critical, however, that these centers be envisioned and planned as
semiautonomous and not simply as ancillary. Downtown Brooklyn is al-

Outer center development possibilities in New York City.

ready one of the largest "central places" in America but continues to be thought of as a back office for Manhattan. The key is zoning for sustainability and difference, not simply for a series of mini-Manhattans. Although the skyscraper is a preeminent symbol of twentieth-century technology and of the culture of the corporation, other paradigms must now emerge as values change. The economic driver that has impelled these heights will be usefully moderated in smaller centers that foreground strong environmental values and in which land prices are restrainedly moderate.

Lower Manhattan is the commercial district with the highest public transportation usage in the country; 80 percent of those who come to work there—350,000 people a day—arrive on mass transit. A comprehensive reexamination and reinforcement of this pattern are crucial to sustaining the city but must be approached noncentrifugally to facilitate movement not simply in and out of Manhattan but between the developing centers of lived life, reinforcing the repatterning. Our waterways, in particular, offer a tremendous opportunity for creating such links with great economy. In addition, the city's large areas of public greenspace and municipally owned property can be used to begin to create a third transport net—for pedestrians, bikers, and nonaggressive zero-emissions vehicles—to supplement the street grid and the subway.

Business as usual in New York City is more than the compulsion to repeat patterns of the past: our talent is creating the new. In the case of downtown Manhattan, however, it is also important to recognize that this is an area of the city that is near completion: its project of build-out and of formal invention is almost done. The construction of the World Trade Center, the isolation of Battery Park City by an overwide highway, the nasty scale of many newer high-rises, the abandonment of the piers, the elimination of manufacturing and small-scale commercial activity, and the elevation of the West Side Highway are all assaults on a satisfying paradigm of great scale contrasts, rich architectural textures, and pedestrian primacy that lies at the core of what's best about Downtown. Restoring this is the task at hand, and it cannot be accomplished in Lower Manhattan alone.

New York, New Deal

IN THE SUMMER OF 1814, facing a threatened British invasion, New York City boiled with activity. Brigades of citizen volunteers—arrayed, as was the custom, by trade, profession, race, and gender—marched off to fashion forts, breastworks, and blockhouses. Artisans and patriotic ladies, lawyers and cartmen, merchants, shopkeepers, and "free people of color" felled trees, dug trenches, and hauled artillery about Brooklyn Heights and Upper Manhattan, while some 23,000 volunteer militiamen, flocking in from the surrounding countryside, drilled and paraded. The crisis passed; that attack never came. But now another one has, of quite a different sort, and contemporary brigades are rallying to the defense and repair of their city. Within weeks of the devastation, the town was abuzz with meetings and conferences. Panels of experts, with large crowds in attendance, joined in passionate discussions about where to go from here. Soon the Internet's virtual agora was festooned with plans for rebuilding the municipality, posted by business groups and unions, architects and planners, churches, communities, and social welfare organizations: citizens wielding word processors rather than muskets.

The most heartening thing about the proposals is their evident desire to make whole—indeed to improve—all of Gotham, not just Ground Zero. September 11 made starkly manifest the interconnecting ties that bind our immensely complicated civic organism. Shock waves juddering out from the blast site set off cascading chains of collateral damage (fear of flying → braked tourism → hotel layoffs → besieged soup kitchens). "Missing" posters in the subways and capsule biographies in the *Times* made clear the distances from which people had come to their fatal downtown rendezvous. Stabbed in Manhattan, we bled in the boroughs and the suburbs, too.

Renewed awareness of and attention to our common weal came just in time: 9/11—and the recession it accompanied and exacerbated—yanked to the front burner a host of problems left too long asimmer. The attack, in making chronic conditions acute, helped galvanize the will to confront them. It also cracked open conventional ways of thinking about *how* to tackle our dilemmas. For nearly a quarter-century now, since the so-called fiscal crisis of the 1970s and Reaganism's subsequent triumph on the national scene, reigning mantra-makers have chanted the ineffectuality, indeed the impermissibility, of purposive public action. In the harsh aftermath of the Twin Towers, the fantasy of privateers—that passive reliance on the "free market" cures all ills—has suddenly come to seem tired, timid, an altogether inadequate response to challenge.

The prying open of such ideological choke holds on public discourse, coupled with the patent urgency of the current crisis and the election of a new mayor and city council, has set long-extant (and newly created) policy organizations lofting programmatic *pdfs* and *html* into hyperspace. The plans' progenitors span the city's sociopolitical spectrum, embracing constituencies whose clashing perspectives and interests have often led to gridlock—real estate developers and welfare rights advocates, stockbrokers and housing reformers, infrastructure builders and environmental activists. Yet to a remarkable degree their proposals overlap or are at least potentially compatible. Indeed, in the aftermath of tragedy, coalitions of these disparate bodies have flowered, their members pledged to work together. This is not to say that everyone is on the same Web page, but rather that I think I can discern, amid the welter of proposals, a roughly congruent vision of what a new New York might look like. Comity, to be sure, could quickly run up against the roadblock of limited resources, setting off battles over priorities. Developers and educators might well agree on the desirability of both offices and schools and yet, in a crunch, insist their own project take precedence on the civic agenda. That is why I want, for just a moment, to set aside the obdurate realities of money and politics, and to assess, in utopian freedom, an inchoate vision of New York's future that I think I see emerging. Once we get a better sense of goals, we can return to the hard work of figuring out how to attain them.

While the shape of a future Ground Zero memorial remains contentious, a much greater degree of convergence marks discussions over the future of Lower Manhattan. The initial desperate desire to shoot up new towers quickly abated once awareness set in that due to relocations and recession, Downtown was now awash in millions of square feet of vacant office space. In similar fashion, the early panicky insistence on doing whatever it took to keep "the financial district" at the island's southern tip gave way to an appreciation that the area long ago lost that distinction, or rather that it came to share it with Midtown and Midtown South, and that today New York's "central business district" is really a decentered multipolar one.

I would argue that Lower Manhattan lost its unchallenged predominance in the 1920s, and that the efforts that began with David Rockefeller's Downtown Lower Manhattan Association in the late 1950s and culminated with the building of the World Trade Center in the 1970s constituted a failed attempt to turn back the clock. After a boom in the mid-1980s, the late eighties–early nineties economic and real estate crash again left Downtown with a whopping vacancy rate. Local and citywide leaders decided to diversify its base, reaching out for new residents by condo-izing old skyscrapers, and for new high-tech businesses by converting empty office space into roosts for new media enterprises. Entrepreneurs arrived; residents moved in (it became the city's fastest-growing locality); and so did some artists and cultural institutions. In the mid- to late 1990s, when the dot-com bubble blew up to fantastic proportions, dreams of preeminence returned—though even during the hoopla about Lower Manhattan's being center of the world, no new towers went up.

Now, while it's agreed that eventually some new Class A office space should rise, there's a chastened agreement that the area should lighten up on dreams of centripetal glory—especially given the new corporate concern for enhanced security via dispersion and redundancy—and accept that the financial sector will be Manhattanwide, with important outriggers in New Jersey, downtown Brooklyn, and Queens's Long Island City. The new vision is the old early nineties vision resurrected—of fostering Downtown's ongoing evolution as a mixed-use site, a 24/7 community with exchanges and brokerage houses as anchors and a hefty component of new high-tech industries, along with newly surfaced retailers who had previously been tucked away in the Twin Towers' subterranean cavern, where they collectively constituted the fifth-largest mall in the United States. The vision somewhat parochially assumes that its sizeable residential constituency will consist overwhelmingly of hip young professionals—Wall Street as Tribecca South—though if the goal is really to create a vibrant live-work environment, it would behoove planners to think about including low-cost housing and services in their mix.

With the pressure to thrust skyward diminished, attention has shifted to upgrading the transportation infrastructure, now widely considered crucial to revivifying Downtown. A host of well-thought-out proposals have emerged for improving the area's links to Midtown, to the rest of the city, and to the surrounding region. Apart from the PATH train from Jersey, there's never been direct commuter access to Lower Manhattan, one reason it lost its primacy to train-terminal-rich Midtown.

There's a larger concern about the financial sector to which 9/11 has drawn attention: wherever it's physically located, New York has become dependent on it to an unhealthy degree. With an ever-greater percentage of our jobs and tax dollars relying on FIRE (finance, insurance, and real estate), we're becoming ever more a one-industry town—problematic for several reasons. One is the tremendous volatility of the money business: when it's hot, it's hot, and bonuses overflow the land; but when it cools, it sheds load rapidly, pulling associated business and information services down with it. Hollywood and Washington may be one-horse towns, but the film and government worlds don't undergo 50 percent swings in earnings. Not so long ago we were less vulnerable to stock market crises. Our economic stool rested on many legs, manufacturing and commerce among them. Now those other legs have been shortened, rotted out, or sawn off, and our situation has become correspondingly perilous.

There's another problem with the nature of our overall economic mix:

its maldistribution of rewards. The financial industry has downsized or exported many of its middle-class positions, leaving behind very highly and very poorly paid employees. Government downsizing has eliminated additional middling positions. So has the decline of manufacturing and commerce jobs, which paid relatively well and afforded climbable career ladders, in large part because they were heavily unionized. Our income distribution has gotten steadily more skewed, and the distance between rich and poor—always substantial throughout New York's history—has grown to Brobdingnagian proportions. It's true that the *number* of jobs here grew vigorously in the 1990s boom—though it did take seven years of recovery and expansion for the city to simply regain all the jobs it lost in the recession of 1989 to 1992. But even with the boom in full swing, the diminution of middle-class positions, coupled with the wildly un-equal reward structures for professionals and managers, on the one hand, and low-level service employees, on the other, generated enormous in-equalities and serious social problems. While the rich did fabulously well, the middle class shrank. And the so-called working poor (their numbers swollen by arriving immigrants and the 350,000 people kicked off wel-fare) became ever more impoverished despite ever more arduous labor, pincered as they were between insufficient wages and escalating expenses. The superheated economy drove up the cost of health care (most working poor had no health insurance). It did the same with housing. The city's population surged by over 450,000 in the 1990s, but only 85,000 new homes were built. Rising rents forced between a quarter and a third of all New Yorkers to spend over half their income on rent, or to burrow into one of the estimated 100,000 illegal (but cheaper) apartments carved out of basements, garages, or subdivided rooms. Lines at soup kitchens and food pantries lengthened even as the boom roared on. By 1999, with irra-tional exuberance at its peak, one of every four New Yorkers lived below the poverty line, a rate twice the national average.

Then came the cruel one-two punch of recession and 9/11. Of the over 100,000 jobs lost, the overwhelming majority were in low-paying service categories. Those who had been living on the edge of a cliff were now pushed off it, sent tumbling towards a shredded safety net. Many of those laid off, particularly former welfare recipients, didn't qualify for unemployment. Many others were legal immigrants who, under the terms of the 1996 welfare "reform" law, were denied access to food stamps. Emergency food service lines exploded, with agencies turning away the hungry in record numbers as supplies ran out. Homelessness jumped to 29,000. New York State—unique in constitutionally guaranteeing aid to the poor—and the federal government stepped in with short-term emergency programs. But a remarkable variety of commentators have come to believe we need a more long-term response, one that diversifies our overall economic structure and adds redundancy by strengthening other components.

Many propose a new overall development strategy, one that goes far beyond our recent reliance on giving tax breaks to big financial and media institutions that threaten to leave town. The two billion dollars in assorted tax breaks we shelled out over the last twelve years no doubt halted some departures. But many of the expensive concessions were granted to industries that were growing rapidly in the nineties, and that clearly considered their Manhattan location vital.

Instead of chasing individual companies with a checkbook, the city should invest in development strategies aimed at manufacturing. Long dazzled by finance, many civic and corporate leaders actively dismissed the production of things as a grungy leftover from the archaic old days. If crummy jobs were leaving for other regions or foreign climes, then good riddance to them. Free-marketeers chimed in with claims that the plummeting number of industrial positions (roughly three quarters of a million from the mid-twentieth-century to now) represented naught but the inevitable (and inevitably benign) consequences of globalization. It's true that the flight of manufacturing was a nationwide phenomenon, and that deep-running forces were partly responsible for the exodus. It's also true, however, that New York City lost such jobs at six times the national rate over the past thirty years—with severe social costs in devastated communities—in part because municipal policy, fixated on big corporate entities, skimped on support to the 96 percent of all city businesses with fewer than fifty employees. This despite the fact that in 1996, these small companies provided nearly three quarters of the city's 2.7 million private-sector jobs, while the targeted big outfits had only 785,000 among them.

Rather than giving $25 million in tax breaks to (maybe) retain 2,000 jobs at one large corporation, some argue we should develop programs that would retain hundreds of firms with twenty employees each, and disburse our incentives to communities throughout the city.

Despite insufficient tending, manufacturing remains a crucial component of the city's economy, with 275,000 jobs remaining (a quite respectable number, especially when set alongside the 490,000 in FIRE). Given the drift away from vast plants to small-scale, just-in-time, batch-production operations, especially in new fields such as biotech, the city is well positioned to do far better than it has. We need to begin with a city-wide analysis of opportunities and then move to targeted sector interventions—we've done Comstat for cops, let's do Jobstat for jobs. We should help specific clusters with R&D support, workforce training, market promotion, export assistance, and the building of support groups such as the Garment Industry Development Corporation, a nonprofit consortium of government, business, and labor founded in 1984 that strengthened the fashion business.

The city could also assist industrial victims of the terrible squeeze on land that developed during the boom, as ad agencies, law firms, and architectural outfits flocked to the Garment District and Chinatown, sending real estate prices soaring and exiling printers and apparel makers (among many other manufacturers) to locations ever more remote from their primary clients. The municipality's Economic Development Corporation helped birth the Greenpoint Manufacturing and Design Center by selling it a city-owned factory building for one dollar, affording space for scores of small woodworkers, designers, and artists, and we could do more along those lines. The city might also establish special industrial districts that combine zoning protections, infrastructure improvements, and industrial development incentives along with pollution controls. It could also facilitate cleanup of the city's roughly 4,000 acres of brownfields, and establish incubators for fledgling high-tech enterprises.

These microtechniques can be applied to biotech—a sector where we really missed the boat—and to a myriad of other small, flexible, light-manufacturing entities geared toward producing high-quality goods for other local businesses—like the bakers who supply hotels and restaurants, the printers who service financial and advertising firms, and the mannequin makers who sell to department stores.

If a touch of Jane Jacobs is required in some quarters, others could use a dash of Robert Moses. As with its manufacturing base, New York let its port facilities slide, then shuffled them (and their jobs) off to Jersey. Again, large economic forces were at work, but were sped along their way by a conviction that go-ahead Manhattan (and even the far more commercially oriented Brooklyn) would be better off without such antediluvian activities. Luckily, opportunity is now knocking once again, and many are urging a swift and positive response. Global trade is increasing exponentially, and its patterns of flow are shifting. Commerce is coming to rely on megacontainer ships, each four football fields long and loaded with upwards of 6,000 giant boxes, and each too big to squeeze through the Panama Canal. It's become cheaper to move a container from Southeast Asia west via Suez across the Atlantic to New York, rather than east across the Pacific to Los Angeles and then by rail to East Coast ports.

Handling these maritime monsters requires blasting and dredging the Kill van Kull to allow them to lumber into the Port Newark and Elizabeth complexes. But even once that very expensive undertaking is completed, New Jersey by itself will be unable to handle the expected explosion in traffic. Proposals have therefore been floated to bring shipping back to Brooklyn (among other deepwater upper harbor locations). Using an approach widely employed in modern European and Asian ports, megaships could dock at offshore concrete caissons—avoiding the need for landfill—where high-speed cranes could transfer containers to barges for direct transhipment around the harbor or to double-stack rail cars at the old 65th Street yards, from where they could be sent over the still-existing tracks of the old Bay Ridge Line, over the Hells Gate Bridge, and on to points north and northeast. Using smart and space-intensive technologies would allow us to revitalize the waterfront commercially, while still preserving great green stretches of it for recreational and residential use.

We also have to sort out our snarled-up land-based transport scene. Right now, containers arrive by rail in Delaware or northern New Jersey and are then transferred to trucks, which, in enormous numbers, traverse the Verrazano or the George Washington Bridge, belching pollutants as they roll through Gotham on their way to Long Island, Westchester, and southern Connecticut—one reason the South Bronx and northern Manhattan have the highest asthma rates in the world. An alternative, long in the works but recently endorsed by a major EDC investment study, calls for construction of a cross-harbor freight tunnel and an increase in the use of rail float cars, both of which would diminish the number of tractor trailers on our roads.

Goods arriving by air need serious attention, too. The aviation industry was badly hurt by 9/11. Queens was devastated by layoffs at JFK and La-Guardia airports and the rippling damage done to freight forwarders, catering companies, limo services, parking lot operators, airport hotels, and bus companies. But our airport infrastructure had been in bad shape before the attacks, having been neglected for years, and while improvements in passenger service are finally under way, the problems of commercial cargo (which accounts for 44 percent of all employment at JFK) remain.

Economic development agendas also include calls for a renewed municipal commitment to providing affordable housing. Nine out of ten business leaders in one survey said the high cost of housing workers was a serious impediment to attracting companies to New York City. The lack of moderate-income shelter makes it hard to recruit municipal workers, too, and drives our teachers and firefighters out of town. Finally, to ameliorate recession-swollen hardships, we need to rebuild and broaden the safety net, easing the way for those entitled to public support rather than strewing their path with obstacles.

How is the city to pay for all these programs? Conventional wisdom says it can't. We closed last year's $1.6 billion budget gap by ruthlessly slashing municipal services, but estimates of this year's shortfall run to four billion dollars. There's "no money," people say. We have to cut back, batten down, tighten belts, wait (hope) for the economic revival that surely lurks just around the corner.

But it isn't true: there's plenty of money around. Our cash flow is low just now, but it's been a mighty flood in recent years, and it's essential to understand how and why it's dwindled. We must recall that a significant portion of our former inflow of income got diverted away from the public treasury into private hands. Over the last decade, we chopped business taxes, repealed a 12.5 percent income tax surcharge imposed during the last recession to enhance public safety, abolished a commuter tax, and gave out $2 billion in assorted tax breaks to would-be runaway companies. Tying off such revenue streams had minimal effect during the boom years, but the cost of all the city's (and state's) tax cuts and retention deals—income that once was ours but now is lost—adds up to the bulk of the current deficit.

We should also stop the drain of revenue to Albany by, for instance, pushing the state to assume the city's Medicaid burden. It would be nice, while we're at it, to stop some of the drain to Washington, D.C. New York pays billions more in taxes than it gets back in grants, contracts, wages, salaries, transfer payments, and all other federal spending. Once the treasury's cracks and crevices have been caulked, it needs to be refilled. We should restore the commuter tax and the stock transfer tax and legislate a modest increase in the real estate tax rate on one-, two-, or three-family houses. We should also, as we have since the Colonial era, impose user fees—charges for the private use of public property. We should start with tolls on East River bridge crossings, adjusting them by time of day (highest in rush hour, lower in off-peak periods, free at night), and by type of user (rebates for carpoolers, intracity commuters, and the small percentage of low-income people who routinely arrive by car). Estimates suggest such fees could generate $700 million a year, a sum that could go either toward general expenses—taking another whack out of the deficit—or strictly to maintenance of roads and bridges and the expansion of mass transit.

In the end, however, undertaking the mammoth projects required for Gotham's revivification will require help from a higher governmental power. We should be making common cause with the millions and millions of people all over the country who are hurting—some from fallout from 9/11, most from the arrival of hard times. Nationwide, virtually every large industry is shrinking; 1.2 million U.S. workers have lost jobs since the recession officially began in March 2001—the biggest drop in twenty years—and another million were forced into the ranks of part-timers. The recession is no respecter of geography. Texas is in trouble, too, particularly Houston, home to Compaq, Continental, and now Enron; San Francisco and San Jose are reeling from the collapse of the silicon bubble, as is Boston; Alton, Illinois, is suffering from setbacks to steel. Across the country, debt-squeezed states and cities are cutting public services.

We should, therefore, immediately strike up alliances with other states and localities and together insist that the federal government (that is, us) should deploy its resources (that is, our tax dollars) to alleviating suffering and revitalizing the economy. We should launch a massive program to create and enhance the nation's social capital, investing in people and resources in a way we haven't done recently but used to do brilliantly.

I'm talking about something far greater than the anemic "stimulus packages" that have been bruited about. We need, I think, a new New Deal.

Looking back at the panoply of federal interventions in the 1930s, three general accomplishments of that distant era seem particularly worthy of emulation. One was the provision of relief in the form of income and jobs for victims of the amoral marketplace. A second was the effort (never completely successful) to jump-start the private economy with a Keynesian jolt of government-underwritten demand. A third was its rehabilitation of the public sector, its marshaling of national resources to augment the nation's social capital. The New Deal and war years created the infrastructure on which much late-twentieth-century prosperity was erected.

In addition to its innovations in energy production, transportation, housing, and regional approaches to problems that transcend state borders, the New Deal initiatives of the thirties that seem particularly relevant to Gotham's current plight are its alphabet agencies (CWA, WPA, PWA), which channeled federal monies to states and localities, allowing them to hire the unemployed and put them to work providing public goods and services. These operations, too, were grounded in ancient New York City responses to crises. As early as the hard times of the winter of 1808, when the port was hobbled by embargo, thousands of unemployed sailors surged through the streets displaying placards demanding "Bread or Work." The city offered both. A municipal soup kitchen was established and seamen lined up for rations three times a week. New York also initiated the country's first work-relief project for those "who are capable of labouring and who are destitute of occupation." The Street Commissioner was directed to hire people to help fill swamps, build streets near Corlear's Hook, lower Murray Hill, and dig the foundation for City Hall. New York, like the nation, would face marketplace collapse over and over again through the centuries, and just as repeatedly would spawn movements demanding remedial government action. The resulting antidepression measures helped build such civic treasures as the Croton Aqueduct and Central Park.

But the nation's most spectacular work-relief programs were those launched by the New Deal. Harry Hopkins, head of the Works Progress Administration, singled out New York City for special attention. Accepting a proposal of Mayor Fiorello LaGuardia, he established a separate WPA unit for the metropolis, treating it as the forty-ninth state. LaGuardia, moving fast, set up a Mayor's Committee on Federal Projects; it put together proposals and won quick approval. By October 1935, New Yorkers were piling onto the federal payroll while other cities were still poring over the application forms. By early 1936, 246,000 were at work on hundreds of white-collar and thousands of engineering projects. The New York City WPA employed more people than any private corporation in town, more people than the War Department. It was one of the biggest enterprises in the United States—a veritable army of labor—and it soon transformed the face of the city.

Roughly two-thirds of WPA employees labored on construction and engineering projects. With astonishing rapidity and efficiency, labor battalions helped build the Triborough Bridge, Lincoln Tunnel, and the Holland Tunnel; extended the West Side Highway and launched the FDR Drive; and constructed LaGuardia Airport, the single most ambitious and expensive WPA undertaking in the nation. In addition, workers repaired and painted fifty bridges, built or rehabbed 2,000 miles of streets and highways (including Queens Boulevard, Jamaica Avenue, and the Grand Concourse from 161st Street to Jerome Avenue), removed 33 miles of trolley tracks, and built boardwalks along Coney Island and Staten Island's south shore. At the same time, it built or fixed 68 piers, laid 48 miles of sewers and 218 miles water mains, erected a host of sewage treatment plants, and conducted pollution control research.

WPA workers also built public amenities that allowed millions of New Yorkers access to benefits not available to them in the prosperous 1920s. The New Dealers refurbished and expanded 287 parks (including Jacob Riis and Mount Morris) and laid out 400 additional ones (including Alley Pond and Cunningham). They built seventeen municipal swimming pools, Orchard Beach in the Bronx, the 20,000-seat Randalls Island stadium, a new zoo in Central Park, and 255 playgrounds in residential neighborhoods. In 1934 alone, they churned out ten municipal golf courses, 240 tennis courts, and 51 baseball diamonds.

To enhance public health care, the WPA built Queens General Hospital, repaired Harlem Hospital, established the city's first clinic to detect and treat outpatients for venereal disease, and started two score baby health stations in dozens of neighborhoods. To ease an education space crisis (classes of forty to fifty students were common), the program renovated and built hundreds of schools, and did major work on Hunter and Brooklyn colleges. Other public buildings erected or improved included public libraries, covered municipal markets, courthouses, homeless shelters, armories, and 391 new firehouses and police stations. In integrated public housing campaigns, the WPA, PWA, and New York City Housing Authority demolished thousands of slum buildings and replaced them with projects like the Williamsburg and Harlem River Houses. And as the nation moved toward World War II, it built or rehabbed barracks and military bases.

The remaining one-third of WPA projects hired out-of-a-job white-collar, service, and professional workers—doctors, nurses, pharmacists, dentists, clerks, typists, housekeepers, orderlies, actors, musicians, and lab technicians—and proceeded to shower residents with novel services. Teachers on the WPA payroll launched adult education classes (in 1938, over 50,000 illiterates, preponderantly recent immigrants, were learning to read), taught inmates in city jails, and developed preprimary schools.

Jobless nurses, doctors, office workers, and dentists staffed nineteen Health Department diagnostic and health care centers around the city, drove about in mobile vans giving X-ray exams that uncovered a thousand cases of active tuberculosis, tested 50,000 for syphilis and gonorrhea, ran diphtheria immunization campaigns, offered dental clinics in schools, did medical research, ran infant care centers, pioneered recreational therapy at Bellevue's childrens' section, and provided household help to the elderly and chronically ill. The project established public day care centers—all but nonexistent before the Depression—in cooperation with the Board of Education. Open from 8:30 to 5:30 to accommodate working mothers, they were geared to preschool children aged two to five, and offered nutritious hot lunches and regular examinations by a registered nurse. Clerical workers organized municipal records, indexed the census, augmented library staffs in nearly every branch, and drove bookmobiles to outlying neighborhoods. Lawyers provided free legal aid to poor litigants.

Encouraged by Eleanor Roosevelt, Hopkins gave work to thousands of unemployed artists, musicians, actors, and writers, declaring: "Hell, they've got to eat just like other people." Muralists painted frescoes in post offices, musicians and vaudevillers gave free concerts and variety shows in city parks. The Federal Theater hired 3,000 actors, dramatists, directors, and stagehands to produce both new and classic plays that reached 30 million people. Writers produced immensely popular guidebooks like the *WPA Guide to New York City*, while the Federal Artists Project supported thousands, among them Stuart Davis, Jackson Pollock, and Berenice Abbot. WPA emergency personnel kept museums functioning (by 1936, they constituted 70 percent of the labor force at the Brooklyn Museum).

The WPA was not perfect, not the be-all and end-all, not something that could or should be mindlessly copied. It was open only to those who could prove destitution after a humiliating inquisition, it did not initially pay prevailing wages and thus undercut union workers; and it discriminated against blacks (FDR felt obliged to kowtow to racist Southern Democrats), though Hopkins made substantial improvements, and African-American leaders appreciated his efforts. Women, too, were shortchanged, partly because the WPA accepted existing gender patterns as givens, and provided them with positions as maids and cosmetologists. The program was, moreover, subject to the ebb and flow of national politics, and its many privateering enemies managed to curtail and eventually kill it. Still, its legacy survived and influenced later initiatives, as when moderate Republicans adopted a variant in the form of revenue sharing. We should support today's mayors and governors in a push to replace tax breaks to corporations and the rich with a massive transfer of federal monies, under reasonable national guidelines, back to badly strapped states and localities. This could be the source of funding for many of the kinds of projects for which New Yorkers are now calling. Getting from here to there will be a tall order, requiring a sea change in national attitudes, but there are substantial grounds for hoping that under the press of hard blows and hard times, a more vigorous response to challenge may yet be forthcoming.

ABOUT THE CONTRIBUTORS

MOUSTAFA BAYOUMI is Associate Professor of English at Brooklyn College, City University of New York. He is coeditor of *The Edward Said Reader* and has published essays in a number of journals, including *Transition*, the *Yale Journal of Criticism*, and the *Journal of Asian American Studies*. Currently, he is completing a manuscript entitled *Migrating Islam: Religion, Colonialism, and Modernity*.

MARSHALL BERMAN is author of *The Politics of Authenticity, All That Is Solid Melts into Air*, and *Adventures in Marxism*. He is currently working on *One Hundred Years of Spectacle: Metamorphoses of Times Square*. He is Distinguished Professor of Political Science at City College and the City University of New York Graduate Center.

M. CHRISTINE BOYER is the William R. Kenan Jr. Professor of Architecture in the School of Architecture, Princeton University. She is a city planner whose interests include the history of cities and the processes of city planning and historic preservation. Her books include *The City of Collective Memory: Its Historical Imagery and Architectural Entertainments Manhattan Manners: Architecture and Style*, and *Dreaming the Rational City: The Myth of American City Planning 1899–1945*.

EDWIN G. BURROWS is Broeklundian Professor of History at Brooklyn College, coauthor with Mike Wallace, of *Gotham: A History of New York City to 1898*, and winner of the Pulitzer Prize in History. Currently, he is writing a book on the memory of the American Revolution in New York City and gathering materials for a history of corruption in the United States.

ERIC DARTON is the author of *Divided We Stand: A Biography of New York's World Trade Center*. He began chronicling the transformation of his native city in the late seventies as art and performance editor for the *East Village Eye*. In recent years Darton has contributed articles on urban culture and institutions to *Designer-Builder*, *Metropolis*, and *Culturefront*. His essays on the cultural history of the Depression-era United States and the rise of European fascism were published in the companion volume to Tim Robbins's award-winning film *The Cradle Will Rock*. Darton's critically acclaimed novel *Free City* was subsequently published in German and Spanish editions. His short-fiction collection, *Radio Tirane*, appeared in *Conjunctions*, where he served as an editor for two years. He is currently fiction editor of *American Letters & Commentary* and a senior contributing editor to the online journal *Frigate*.

KELLER EASTERLING is an architect, writer, and associate professor at Yale. Her recent book, *Organization Space: Landscapes, Highways and Houses in America*, applies intelligence from information technologies to a discussion of American infrastructure and development formats. She is also author of *Call It Home*, a laser-disc history of American suburbia from 1934 to 1960. She is currently working on a book entitled *Terra Incognita: Pirate Space in Global Development*, which is about "U.S.-style" spatial formats exported to pivotal political locations around the world.

BEVERLY GAGE is a historian and journalist currently completing her Ph.D. in U.S. history at Columbia University. Her work has appeared in *The Nation* and Salon.com, among other publications. Her book on the 1920 Wall Street explosion will be published by Oxford University Press in 2003.

DAVID HARVEY is Distinguished Professor of Anthropology at the City University of New York Graduate Center. He was previously Professor of Geography at the Johns Hopkins University and at Oxford University. He is the author of numerous books, including *Social Justice and the City*, *The Limits to Capital*, *The Condition of Postmodernity*, *The Urban Experience*, and *Spaces of Capital*.

SETHA M. LOW is Professor of Environmental Psychology and Anthropology at the Graduate Center of the City University of New York. Her books include *The Anthropology of Space and Place* (with D. Lawrence), *On the Plaza*, *Theorizing the City*, *Children of the Urban Poor* (with F. J. Johnston), *Place Attachment* (with I. Altman), and *Housing, Culture, and Design* (with E. Chambers).

PETER MARCUSE, a lawyer and urban planner, is Professor of Urban Planning at Columbia University in New York City. His many works include *Globalizing Cities* (coedited with Ronald van Kempen) and *Of States and Cities* (coedited with Ronald Van Kempen). He is on the editorial boards of a number of professional journals, and has been a consultant to local, state, and national government on housing policy issues.

ROBERT PAASWELL currently serves as Director of the federally supported University Transportation Research Center, located at the City College of New York. He is also Director of the City University Institute for Urban Systems, a major university-wide initiative to examine the intersection of new technology, changing institutional structure, and innovative finance on the provision of infrastructure in the twenty-first century. He is a Distinguished Professor of Civil Engineering and served as Executive Director (CEO) of the Chicago Transit Authority.

ANDREW ROSS is Professor and Director of the Graduate Program in American Studies at New York University. His books include *The Celebration Chronicles: Life, Liberty and the Pursuit of Property Value in Disney's New Town; Real Love: In Pursuit of Cultural Justice ; The Chicago Gangster Theory of Life: Nature's Debt to Society; Strange Weather: Culture, Science and Technology in the Age of Limits*; and *No Respect: Intellectuals and Popular Culture*. He has also edited several books, including *No Sweat: Fashion, Free Trade, and the Rights of Garment Workers*. His book on modern work, entitled *No-Collar: The Humane Workplace and its Hidden Costs*, is forthcoming from Basic Books.

ARTURO IGNACIO SÁNCHEZ teaches planning at Pratt University. He writes regularly about Latino demographic and immigrant issues for *HOY*, the largest Spanish-language daily newspaper in the tristate area. He is also a member of New York City's Community Board 3 in Queens, which encompasses Jackson Heights, Corona, and East Elmhurst.

NEIL SMITH is Distinguished Professor of Anthropology at the City University of New York Graduate Center, where he also directs the Center for Place, Culture and Politics. He is the author of *The New Urban Frontier: Gentrification and the Revanchist City; Uneven Development: Nature, Capital and the Production of Space*, and *The Geographical Pivot of History: Isaiah Bowman and the Geography of the American Century.*

MICHAEL SORKIN is the principal of the Michael Sorkin Studio in New York City and is the Director of the Graduate Urban Design Program at the City College of New York. Recent projects include master-planning in Hamburg and Schwerin, Germany, planning for a Palestinian capital in East Jerusalem, campus planning at the University of Chicago, and studies of the Manhattan waterfront and Arverne, Queens.

His books include *Variations on a Theme Park, Exquisite Corpse, Local Code, Wiggle, Some Assembly Required, Other Plans*, and *The Next Jerusalem*.

JOHN KUO WEI TCHEN is Associate Professor of History at New York University, where he is also Director of the Asian/Pacific/American Studies Program. He is the author of *New York before Chinatown: Orientalism and Shaping of American Culture, 1776–1882* and editor of *Genthe's Photographs of San Francisco's Old Chinatown.*

MIKE WALLACE is Distinguished Professor of History at John Jay College and the City University of New York Graduate Center. He is the recipient of the Pulitzer Prize in History for the critically acclaimed *Gotham: A History of New York City to 1898*, which he coauthored with Edwin G. Burrows, and founder of the Gotham History Center.

MARK WIGLEY is a professor at the Columbia School of Architecture. He is the author of *White Walls, Designer Dresses, The Architecture of Deconstruction*, and *Constant's New Babylon*.

SHARON ZUKIN is Broeklundian Professor of Sociology at Brooklyn College and the City University Graduate Center in New York. Her books include *Loft Living, Landscapes of Power: From Detroit to Disney World*, and *The Cultures of Cities* as well as *Beyond Marx and Tito* and the edited volumes *Structures of Capital* (with Paul DiMaggio) and *Industrial Policy*. She has written extensively about culture and economic change in cities, especially in New York, and is now writing a book about shopping and cultures of consumption. *Landscapes of Power* won the C. Wright Mills Award of the Society for the Study of Social Problems.

INDEX